# TRENDS IN DNA RESEARCH

# TRENDS IN DNA RESEARCH

# TRENDS IN DNA RESEARCH

COREY R. WOODS
EDITOR

Nova Biomedical Books
*New York*

For permission to use material from this book please contact us:
Telephone 631-231-7269; Fax 631-231-8175
Web Site: http://www.novapublishers.com



**Library of Congress Cataloging-in-Publication Data**
*Available upon request*

ISBN 1-59454-635-5

*Published by Nova Science Publishers, Inc. ✣New York*

# Contents

# Preface

Deoxyribonucleic acid (DNA) is a chemical found primarily in the nucleus of cells. DNA is a long, spiralling molecule that orchestrates the cell's daily operations and provides the genetic blueprint for the physical characteristics of all living organisms. It is the molecule that encodes genetic information in the nucleus of cells. It determines the structure, function and behaviour of the cell. DNA is made up of two complementary strands, the strands intertwine like a spiral staircase to form a structure called a double helix. Subunits, called bases, are the rungs of the staircase. The four nucleotides in DNA contain the bases: adenine (A), guanine (G), cytosine (C), and thymine (T). This new book presents leading-edge research in this dynamic field.

In chapter I, inflammatory responses are essential defense mechanisms which protect pluricellular organisms from various types of insult and imply the regulation of a vast number of genes. The transcription factors NF-κB and AP-1 are key players in inflammation, since they are rapidly activated by inflammatory stimuli and in turn direct the expression of inflammatory mediators such as cytokines, matrix metalloproteinases, iNOS and COX-2. The inflammatory response is a tightly controlled process. It possesses intrinsic mechanisms that limit its intensity and duration. Key inflammatory mediators such as nitric oxide and prostaglandins may contribute both to the onset and to the resolution of inflammation. Cyclopentenone prostaglandins (cyPG) are reactive compounds which may form covalent adducts with thiol groups in proteins or in glutathione. CyPG have proven potent antiinflammatory agents both in cellular and in animal models. CyPG may act as ligands for the nuclear receptors of the PPAR family of transcription factors, and PPAR-mediated responses appear to contribute to their beneficial effects. In addition, posttranslational modification of proteins by attachment of cyPG may modulate signal transduction cascades leading to transcription factor regulation and/or affect the activity of transcription factors themselves. The transcription factors NF-κB and AP-1 have been identified as targets for this modification. In this work we have explored the effects of cyPG on the NF-κB activation pathway in mesangial cells, a known cellular model of inflammation. Activation of mesangial cells in the presence of the cyPG 15-deoxy-$\Delta^{12,14}$-prostaglandin $J_2$ reduces NF-κB transcriptional activity. Pre-treatment with cyPG inhibits NF-κB DNA binding activity both in intact cells and in isolated nuclear extracts. By using a biotinylated 15-deoxy-$\Delta^{12,14}$-prostaglandin $J_2$ derivative we have observed the modification of the NF-κB subunits p65 and p50 in intact cells. Subsequently, we have used biotinylated cyPG to study the interaction of these prostanoids with NF-κB *in vitro*. Binding of biotinylated cyPG to p50 is selectively reduced in the presence of an oligonucleotide containing the consensus site for NF-κB binding, which suggests that this interaction involves the DNA binding domain. Labeling studies of a cys62ser p50 mutant confirm that this cysteine residue is a site for cyPG attachment. Biotinylated cyPG labeling assays indicate that other proinflammatory transcription factors are potential targets for modification by cyPG. Given the fact that a number of endogenous lipids as well as natural products possess reactive unsaturated carbonyl groups, the approaches used in this study could prove useful to assess their potential in the regulation of transcription factor activity.

As reported in chapter II, Mitochondrial DNA 5178 cytosine/adenine (Mt5178 C/A) polymorphism, which is also known as NADH dehydrogenase subunit-2 237 leucine/methionine (ND2-237 Leu/Met) polymorphism, is reportedly associated with longevity in the Japanese population, and the Mt5178A genotype may possess antiatherogenic properties. Individuals with the Mt5178A genotype may be more resistant to atherogenic diseases, such as myocardial infarction and cerebrovascular diseases, than those with Mt5178C. Moreover, Mt5178 C/A polymorphism has been reported to be associated with blood pressure, intraocular pressure, serum protein fraction levels, and serum lipid levels in healthy Japanese individuals.

In order to investigate the interaction between Mt5178 C/A polymorphism and habitual smoking on serum lipid levels, the authors performed a cross-sectional study in which healthy Japanese male subjects were classified into two groups based on body mass index (BMI): BMI $\leq$ 22 (n = 107); and BMI > 22 (n = 214). Interaction between Mt5178 C/A polymorphism and cigarette smoking was observed with regard to serum triglyceride (TG) levels in BMI $\leq$ 22 subjects ($P$ = 0.026), and with regard to serum total cholesterol (TC) levels and serum low-density lipoprotein cholesterol (LDLC) levels in BMI > 22 subjects ($P$ = 0.019 and $P$ = 0.043, respectively). In BMI > 22 subjects, TC levels were significantly lower in non-smokers with Mt5178A than in smokers with Mt5178A ($P$ = 0.015), while LDLC levels were lower in non-smokers with Mt5178A than in those with Mt5178C or in smokers with Mt5178A ($P$ = 0.047 and $P$ = 0.055, respectively). Although interaction between Mt5178 C/A polymorphism and habitual smoking was not observed with regard to TG levels in BMI > 22 subjects ($P$ = 0.308), TG levels were significantly higher in smokers with Mt5178A than in non-smokers with Mt5178A as well as in non-smokers with Mt5178C ($P$ = 0.011 and $P$ = 0.014, respectively). Taken together, these results suggest that habitual smoking may deprive Japanese Mt5178A (ND2-237Met) genotypic men having BMI of > 22 of their antiatherogenic advantage.

Chapter III focuses on, in order to identify prognostic factors regarding tumor diseases, which consequently determines the individual approach of therapy, it is necessary to first characterize the biological potential of the cancer. Standard methods in the diagnosis and prognosis of prostate and bladder carcinomas provide only limited information on the tumorbiological behavior. At present, the practical role of DNA cytophotometry is emerging. Apart from stage and tumor differentiation, it presents an independent criterion in the diagnosis of tumor rating. Furthermore, DNA cytophotometry helps to evaluate biological tumor progression and the consequent therapeutic approach. In carcinoma of the prostate the authors find it inevitable to determine the degree of DNA malignancy before therapy planning takes place. In addition, DNA cytometry evaluates hormone sensitivity of the tumor cells, especially in cases of heterogeinity of the prostate cancer. By using the stemline interpretation method of Fu they obtained information about the differential diagnosis and evaluation of the progressive character of prostate tumors. In regard to bladder carcinoma, the DNA image cytometry seems to be a prognostical tool in understanding the tumor biological behavior. Estimating of the risk of recurrence in patients after transurethral resection is possible by means of static DNA cytometry. Hence it may provide a valid criterion in the selection of adjuvant chemotherapy. But in the diagnosis of bladder cancer, the results and validity of regular clinical use of DNA image cytometry are controversial. For receiving the highest sensitivity of the measurement a combination of various DNA parameters is helpful, as could be demonstrated by their own scheme. Despite all criticisms and disadvantages, DNA image cytometry should be one of the tools of urological oncology, especially in diagnosing and for prognostic statements of prostate and bladder tumors.

Chapter IV informs about, that it was shown that the cortisol-apolipoprotein A-I complex does not affect the rate of DNA and protein biosynthesis in primary culture of hepatocytes, while the tetrahydrocortisol-apolipoprotein A-I complex (THC-apoA-I) considerably accelerates the insertion of $^{3}$H-thymidine into DNA and $^{14}$C-leucine into protein. Molecular mechanism of this phenomenon was studied. The in vitro action of the apolipoprotein A-I-tetrahydrocortisol complex on a native DNA isolated from mature rat liver was shown to cause the formation of S1-nuclease-sensitive regions. The most interactive sites were distributed over the genome irregularly. After the interaction with tetrahydrocortisol-apolipoprotein A-I complex, DNA gains the capacity of dose-dependent and highly cooperative binding to RNA-polymerase. The reduced form of hormone, tetrahydrocortisol, that was previously considered to be an inactive metabolite, when being composed with the complex with apolipoprotein A-I gains ability to carry out the biological function similar to the function of a transcriptional factor. Using the method of small-angle X-ray scattering (SAXS) it was demonstrated that a pronounced interaction with eukaryotic DNA is observed only in the case of THC-apoA-I complex. This interaction is accompanied by local melting of DNA. The most probable site of binding this complex to DNA is a sequence of $(GCC)_n$ type composing various eukaryotic genes, human genes as well. Oligonucleotide (duplex) of this type has been synthesized. Its interaction with THC-apoA-I complex was shown to cause the formation of a more composite complex, which dissociates into complementary oligonucleotides. The latter ones also interact with THC-apoA-I complex. The kinetics of this multistep process is presented. Interaction of the cortisol-apolipoprotein A-I complex with the duplex is less specific and does not lead to the formation of complementary oligonucleotides. Study of the mechanism of interaction between duplex $CC(GCC)_5 \cdot GG(CGG)_5Li_2$ and THC or THC-apoA-I complex revealed the formation of hydrogen bonds between OH group of the THC A-ring and C=O group of cytosine or guanine. Interaction with cortisol or cortisol-apoA-I complex leads to the formation of hydrogen bonds with NH group of cytosine. In addition, THC and cortisol form hydrogen bonds with $PO_2$ group of the duplex and with OH group of monosaccharide. Interaction of apoA-I with the duplex is accompanied by the formation of hydrogen bonds with $NH_2$ group of protein, with C=O group of cytosine, and P=O group of desoxyribose phosphate.

Interaction of THC, apoA-I and THC-apoA-I complex with the duplex occurs at the same active site, namely, with C=O group. The order → disorder structural transition takes place in the duplex under the action of apoA-I or THC-apoA-I and cortisol-apoA-I complexes. Thus, it was shown for the first time that reduced forms of steroid hormones have a significant biological effect. In complex with apolipoprotein A-I they enhance the rate of protein biosynthesis in parenchymal cells. The reported mechanism underlies intracellular regeneration.

Chapter V presents D-Amino-acid oxidase (DAO) which catalyzes oxidative deamination of D-amino acids was found nearly 70 years ago in mammals. However, since D-amino acids are considered very rare in mammals, the physiological role of DAO has been enigmatic. Development of analytical methods has revealed that D-amino acids are present in considerable amounts even in mammals. Therefore, it becomes clear that the metabolism of these D-amino acids is one of the functions of DAO. In addition, brain DAO is shown to have a function to modulate neurotransmission. D-Serine in the brain is shown to bind to the *N*-methyl-D-aspartate (NMDA) subtype glutamate receptors to potentiate the neurotransmission. By catabolizing this D-serine, DAO is considered to further regulate the neurotransmission. The origin of D-serine was not clear, but serine racemase which converts L-serine to D-isomer has been finally discovered in the brains of rodents and humans. Recently, DAO is shown to be implicated in schizophrenia. The newly discovered gene product activates DAO and reduces D-serine, resulting in the hypofunction of NMDA receptors. This hypofunction is likely a cause of schizophrenia. These findings have been resolving the enigma which lasts over 70 years.

As reported in chapter VI, the rapid growth of DNA sequences data in various DNA databanks leads to that the sequence analysis becomes more important but still difficult task. To extract the information stored in the sequence consisting of letters on the alphabet $\Sigma = \{A, C, G, T\}$, many mathematical schemes have been proposed. They are mainly the graphical representations, the numerical characterizations, and the method of characteristic sequences. Among them, the graphical representation gives a visual and hence qualitative characterization of DNA sequences, while the numerical characterization provides a quantitative analysis of data on DNA sequences. The method of characteristic sequences is a new approach based on the idea of the coarse-grained description of the DNA primary sequence.

In chapter VII, the impact of three polymeric surfaces, i.e. carboxy-rich Poly(Styrene/Maleic Acid) (PSMA) and plasma treated cyclo-olefin-copolymer (COC) and polycarbonate (PC), on immobilization efficiency of oligonucleotides has been analyzed. The employment of different immobilization strategies, either chemically or via poly(L-lysine), have resulted in comparable high density (up to $16 \times 10^{12}$ molecules/cm$^2$) of amino-terminated oligonucleotides covalently immobilized on polymeric surfaces. The nature of DNA - poly(L-lysine) complexation is discussed. The relatively high immobilization efficiency coupled with the stable thermo-mechanical properties makes these polymers promising candidates for DNA-DNA hybridization assays integrated with PCR applications.

As reported in chapter VIII, known exemplar samples of human DNA have traditionally been body fluids, such as blood and saliva. In each case, the presence of water increases the risk for bacterial growth which may degrade the DNA evidence. In this study, the authors have developed a method that employed a hydrophilic adhesive tape (HAT) for collecting DNA evidence. The HAT method was used to remove surface cells from relatively hairless areas on the body. The areas examined were ankle, arm, behind the ear, between fingers and back of the neck. The HAT was then dissolved in the extraction buffer. DNA typing was performed using the short tandem repeat (STR) analysis. Our results show that the samples collected from the ear give the best results with a success rate of 100%. All subjects tested by this method had known STR genotypes established from buccal swabs. The author's results suggest that the HAT method can be used as a less invasive method for collecting biological evidence for forensic DNA analysis. In addition, this collection method should reduce the risk of DNA degradation due to the moisture.

In: Trends in DNA Research
Editor: Corey R. Woods, pp. 1-31

ISBN 1-59454-635-5


*Chapter I*

# Regulation of Proinflammatory Transcription Factors by Direct Modification with Cyclopentenone Prostaglandins

***Eva Cernuda-Morollón and Dolores Pérez-Sala****

Departamento de Estructura y Función de Proteínas. Centro de Investigaciones Biológicas. Consejo Superior de Investigaciones Científicas (C.S.I.C.). Madrid, Spain

## Abstract

Inflammatory responses are essential defense mechanisms which protect pluricellular organisms from various types of insult and imply the regulation of a vast number of genes. The transcription factors NF-κB and AP-1 are key players in inflammation, since they are rapidly activated by inflammatory stimuli and in turn direct the expression of inflammatory mediators such as cytokines, matrix metalloproteinases, iNOS and COX-2. The inflammatory response is a tightly controlled process. It possesses intrinsic mechanisms that limit its intensity and duration. Key inflammatory mediators such as nitric oxide and prostaglandins may contribute both to the onset and to the resolution of inflammation. Cyclopentenone prostaglandins (cyPG) are reactive compounds which may form covalent adducts with thiol groups in proteins or in glutathione. CyPG have proven potent antiinflammatory agents both in cellular and in animal models. CyPG may act as ligands for the nuclear receptors of the PPAR family of transcription factors, and PPAR-mediated responses appear to contribute to their beneficial effects. In addition, posttranslational modification of proteins by attachment of cyPG may modulate signal

* Address for correspondence: Dr. Dolores Pérez-Sala. Departamento de Estructura y Función de Proteínas, Centro de Investigaciones Biológicas, C.S.I.C., Ramiro de Maeztu, 9, 28040 Madrid, Spain. Phone: 34 918373112; Fax: 34 915360432; email: dperezsala@cib.csic.es

transduction cascades leading to transcription factor regulation and/or affect the activity of transcription factors themselves. The transcription factors NF-κB and AP-1 have been identified as targets for this modification. In this work we have explored the effects of cyPG on the NF-κB activation pathway in mesangial cells, a known cellular model of inflammation. Activation of mesangial cells in the presence of the cyPG 15-deoxy-Δ12,14-prostaglandin J2 reduces NF-κB transcriptional activity. Pre-treatment with cyPG inhibits NF-κB DNA binding activity both in intact cells and in isolated nuclear extracts. By using a biotinylated 15-deoxy-Δ12,14-prostaglandin J2 derivative we have observed the modification of the NF-κB subunits p65 and p50 in intact cells. Subsequently, we have used biotinylated cyPG to study the interaction of these prostanoids with NF-κB in vitro. Binding of biotinylated cyPG to p50 is selectively reduced in the presence of an oligonucleotide containing the consensus site for NF-κB binding, which suggests that this interaction involves the DNA binding domain. Labeling studies of a cys62ser p50 mutant confirm that this cysteine residue is a site for cyPG attachment. Biotinylated cyPG labeling assays indicate that other proinflammatory transcription factors are potential targets for modification by cyPG. Given the fact that a number of endogenous lipids as well as natural products possess reactive unsaturated carbonyl groups, the approaches used in this study could prove useful to assess their potential in the regulation of transcription factor activity.

## Abbreviations

| | |
|---|---|
| PG | prostaglandin(s); |
| CyPG | cyclopentenone prostaglandin(s); |
| COX | cyclooxygenase; |
| 15d-$PGJ_2$ | 15-deoxy-$\Delta^{12,14}$-$PGJ_2$; |
| NF-κB | nuclear factor-kappaB; |
| IKK | IkappaB kinase; |
| PPAR | peroxisome proliferator activated receptor; |
| AP-1 | activator protein 1; |
| STAT-1 | signal transducer and activator of transcription 1; |
| CREB | cAMP response element binding protein; |
| iNOS | inducible nitric oxide synthase; |
| NO | nitric oxide; |
| GSH | reduced glutathione; |
| HIF | hypoxia inducible factor; |
| RMC | rat mesangial cells; |
| HRP | horseradish peroxidase; |
| Ck | cytokines; |
| IL-1β | interleukin 1-beta; |
| TNF-α | tumor necrosis factor alpha; |
| LPS | bacterial lipopolysaccharide; |

| | |
|---|---|
| EMSA | electrophoretic mobility shift assay; |
| PMA | phorbol 12-myristate 13-acetate; |
| ECL | enhanced chemiluminiscence. |

# Introduction

The inflammatory response is a complex process by which organisms react to trauma or infection. It implies the interaction between multiple soluble factors and many types of cells in a tightly controlled way [1]. Tissue injury induces the release of bioactive peptides from neurons, and of intracellular products from damaged cells. Immune cells present in tissues and leukocytes recruited from the blood vessels release cytokines, eicosanoids and histamine that contribute to vasodilatation and edema [1]. Consequently other mediators are activated, including matrix metalloproteinases and the systems involved in the generation of reactive oxygen and nitrogen species, such as superoxide anion and nitric oxide (NO). The aim of this rapid response is to kill invasive bacteria and destroy damaged tissue, but for this process to lead to healing it is necessary that negative control mechanisms are in place to avoid excessive damage or the development of a chronic situation. Several inflammatory mediators display a dual role that can either promote or suppress inflammation depending on their concentration or the activation state of the target cells.

NO and prostaglandins (PG) are key mediators of the inflammatory process. In inflammatory settings the main source of NO is the inducible form of nitric oxide synthase or iNOS. iNOS is not normally expressed in most tissues but its levels increase rapidly in response to proinflammatory agents such as cytokines or bacterial lipopolysaccharide (LPS). iNOS expression is tightly controlled by transcriptional and posttranscriptional mechanisms [2]. iNOS can generate high amounts of NO, a free radical which may directly modulate protein function by modifying specific residues. NO can also regulate the activity of enzymes such as cyclooxygenases and soluble guanylate cyclase through the interaction with the heme group of these enzymes. The activation of soluble guanylate cyclase leads to an increased generation of cGMP, one of the main mediators of NO action. NO plays a multifunctional role in inflammation [3]. Among the proinflammatory actions of NO are the induction of vasodilation and the stimulation of the expression of proinflammatory genes. It has been observed that NO can also contribute to the resolution of inflammation through the induction of macrophage apoptosis, and the regulation of gene expression [4].

PG released in inflammatory situations contribute to vasodilatation, edema and fever. Therefore PG mediate many of the symptoms of acute and chronic inflammation. PG are synthesized from arachidonic acid by several enzymatic steps. Arachidonic acid is released from membrane phospholipids by the action of phospholipases. The induction of cyclooxygenase-2 expression, the rate limiting enzyme in the transformation of arachidonic acid into PG, plays a key role in the increase of PG associated with inflammatory situations. COX-2 catalyzes the formation of $PGH_2$, which is in turn converted by isomerases into the different PG in a tissue- or cell type-related fashion. In addition to enzymatic synthesis, several types of PG can arise from the non-enzymatic transformation of arachidonic acid or of other prostaglandins.

The interaction between iNOS and COX-2 pathways plays a central role in the regulation of inflammation. The expression of iNOS and COX-2 appears to be coordinately regulated in many settings. Numerous proinflammatory stimuli trigger the expression of both enzymes. In addition, this induction can be subjected to the modulatory action of antiinflammatory cytokines, enzyme cofactors, corticosteroids and reactive oxygen species, which influence the levels of both proteins in a similar way [5-7]. We and others have explored the mechanisms underlying the interplay between iNOS and COX-2 induction in a cellular model of inflammation such as renal glomerular mesangial cells (MC) exposed to proinflammatory stimuli [8-11]. MC are specialized smooth muscle cells which are located around the capillaries of the renal glomerulus where they play an important role in the maintenance of glomerular structure and regulation of glomerular filtration rate [12, 13]. Upon exposure to inflammatory stimuli, these cells suffer a trans-differentiation process to an activated phenotype and become capable of releasing inflammatory mediators such as NO, eicosanoids, chemokines and cytokines. Activated MC also express integrins and adhesion molecules which participate in the recruitment and activation of macrophages. MC express iNOS and COX-2 in response to proinflammatory agents, which are responsible for an increased generation of NO and prostaglandins [14, 15].

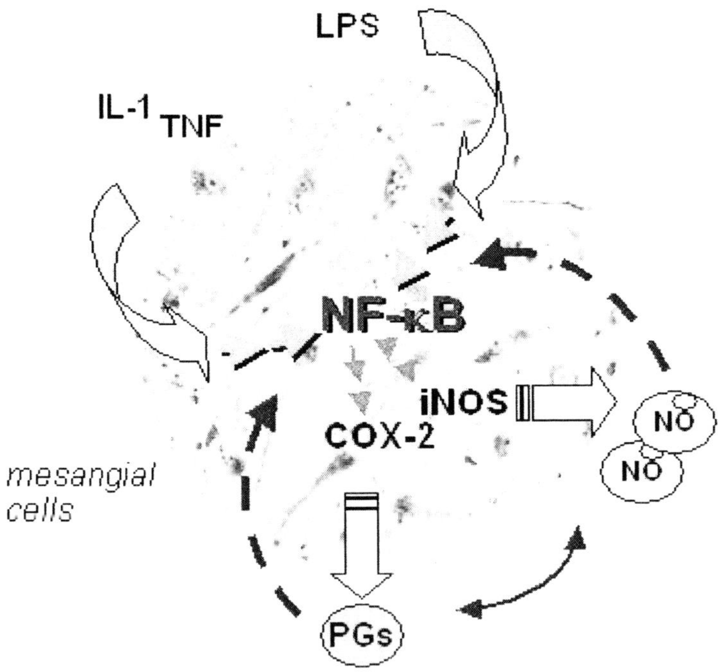


Figure 1. The interdependence of iNOS, COX-2 and NF-κB signaling pathways plays a key role in the regulation of the inflammatory response in mesangial cells. Proinflammatory stimuli trigger the expression of iNOS and COX-2 in mesangial cells. The NO and PG biosynthetic pathways are subjected to significant cross-talk since their products may regulate the activity and/or the expression of iNOS and COX-2. The transcription factor NF-κB mediates the induction of iNOS and COX-2 by cytokines and is an important target for the regulatory effects of NO and PGs on gene expression.

When analyzing the cross-talk between the NOS and COX pathways it is important to consider that the products of these enzymes may exert a feedback regulation of both pathways, as shown schematically in Figure 1. NO has been reported to modulate COX activity both *in vitro* and *in vivo* [16-18]. NO may also modulate COX-2 expression. The effects of NO on the levels of COX-2 are complex and depend on the cell type, concentration, and time. Both amplifying and inhibitory effects of NO on COX-2 expression have been reported in diverse experimental settings [19-21]. We have previously described that NO donors can exert a biphasic regulation of COX-2 levels in mesangial cells [11]. NO also modulates iNOS levels through multiple mechanisms. This modulation can also follow a biphasic pattern [22]. We have postulated that this type of effect may contribute to limit the amplitude or the duration of the inflammatory response. Interestingly, COX products may modulate both iNOS and COX-2 expression. In the early stages of inflammation, COX products may contribute to an amplification of the induction of these proteins, while at later instances or when they reach higher levels they can display an inhibitory effect [23]. A switch from a proinflammatory to an antiinflammatory nature and/or role of the eicosanoids generated during the inflammatory process has been reported that will contribute to the onset and the resolution of inflammation [24, 25].

The molecular mechanisms by which this intricate regulation takes place occur at multiple levels, including the interaction with membrane receptors or intracellular proteins, the generation of second messengers of NO such as cGMP and of PG like cAMP. An important level of regulation involves the modulation of transcription factor activity. Many transcription factors contribute to the inflammatory response. Among the most extensively studied are AP-1 and NF-κB. The transcription factor AP-1 is formed by homo or heterodimers of proteins of the Jun and Fos families. AP-1 is activated in response to cytokines, mitogenic signals and cellular stress [26]. Regulatory sites for the binding of AP-1 are present in the promoters of numerous proinflammatory genes, including iNOS, COX-2 and matrix metalloproteinases. The transcription factor NF-κB is considered a key player in inflammation. NF-κB is a dimeric factor formed by members of the Rel family of proteins [27]. The prototypical member of the NF-κB family is a heterodimer consisting of p65/RelA and p50/NF-κB1. Both subunits possess DNA binding activity, but only the p65 subunit has transactivating activity. In a very simplified view, stimuli such as cytokines and mitogenic signals initiate a cascade of events that leads to the activation of a kinase, known as IKK, which specifically phosphorylates IκB, the inhibitory subunit of NF-κB [28]. This triggers the ubiquitination and degradation of IκB by the proteasome and favors the accumulation of the active NF-κB dimers in the cell nucleus where the factor binds to sites in the promoters of many genes involved in inflammation such as iNOS, COX-2, intercellular adhesion molecule-1 (ICAM-1) and monocyte chemoattractant protein-1 (MCP-1), among others. Mutation of these sites frequently results in a drastic reduction in promoter inducibility by proinflammatory stimuli [7]. As schematized in Figure 1, NO and PG exert an important feedback regulation of NF-κB. NO can bring about both positive and negative effects on NF-κB activity. NO can activate NF-κB by a Ras-mediated mechanism depending on the direct modification of this signaling protein by S-nitrosylation at cysteine 118. A NO-dependent amplification of NF-κB activation by proinflammatory stimuli has also been observed, which has been shown to correlate with a delay in the resynthesis of IκB [11]. The generation of elevated levels of NO when the inflammatory response is established may exert negative effects on NF-κB activation, and this in turn may constitute an important mechanism contributing to the resolution of inflammation. In fact, the effect of genetic or pharmacological suppression of iNOS may lead to biphasic effects on the inflammatory response [22, 29, 30]. The inhibition of NF-κB by NO can occur by cGMP-dependent and independent mechanisms. NO can amplify IκB expression and interfere with the phosphorylation and degradation of IκB in a cGMP-independent manner [31, 32]. Recently, a mechanism for the direct inhibition of NF-κB DNA binding by NO has been characterized. NO can directly S-nitrosylate a cysteine residue located in the DNA binding domain of the NF-κB subunit p50, cysteine 62 in human p50, which needs to be in a reduced state to allow efficient DNA binding [33]. Moreover, in the presence of physiological concentrations of glutathione, NO donors can induce the formation of a mixed disulfide between glutathione and p50 which involves cysteine 62 [34]. In addition, NO may induce other oxidative modifications of NF-κB proteins that may contribute to the inhibitory effect [34].

NF-κB activity is also regulated by prostanoids. PG interacting with membrane receptors can lead to an increase in cAMP concentrations. cAMP in turn can modulate NF-κB activation by several mechanisms, including the phosphorylation of IκB and of the NF-κB subunit p65 by cAMP-dependent protein kinase, which leads to increased activation, which can be dependent in part on the recruitment of coactivators [35, 36]. A direct increase in the transactivation potential of p65 without changes in nuclear accumulation has been invoked for the potentiation of TNF-elicited NF-κB activation by PGE or PGE analogs [37]. In addition, increases in cAMP triggered by treatment with PGE have been shown to correlate with an induction of the mRNA of p65 [38]. In some experimental settings, $PGE_2$ can inhibit NF-κB activation [39]. The nature of the modulation of the inflammatory response by $PGE_2$ may thus depend on the cell type or the nature of the co-stimulatory signal. Recently, much attention has been devoted to a type of PG known as cyclopentenone prostaglandins (cyPG) (Figure 2A).

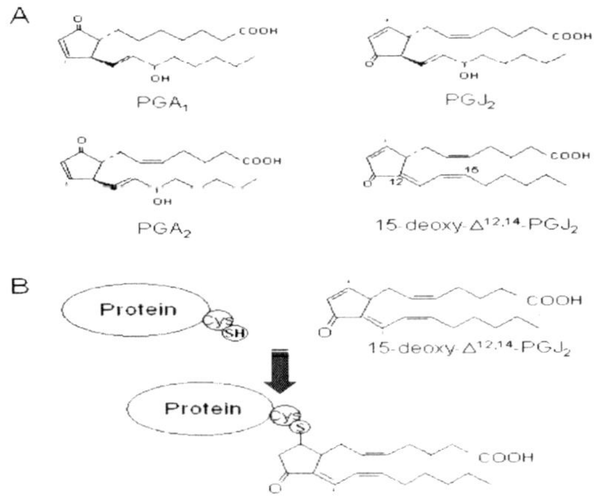


Figure 2. Structure of cyPG and formation of Michael adducts with proteins. (A) Structures of several cyPG of the A series (left) and of the J series (right). Electrophilic carbons are marked by asterisks. (B) Schematic representation of the formation of a Michael adduct between the electrophilic carbon in the cyclopentenone ring of 15d-$PGJ_2$ and the sulphydryl group of a protein cysteine residue.

These endogenous eicosanoids are generated by the spontaneous dehydration of their parent PG. More precisely, PG of the A series arise from the dehydration of PGE, while PG of the J series are produced by dehydration of PGD. These PG owe their name to the presence in their structure of an α,β-unsaturated carbonyl group in the cyclopentane ring (cyclopentenone). CyPG have been detected in biological fluids and tissues in significant amounts. Although the magnitude of the generation and the bioavailability of these compounds is still a matter of controversy, increases in cyPG generation have been detected in association with COX-2 induction in several pathophysiological situations. The cyPG 15-deoxy-$\Delta^{12,14}$-$PGJ_2$ (15d-$PGJ_2$) has been detected in the foamy macrophages of human atherosclerotic plaques [40] and in spinal cord sections of patients with sporadic amyotrophic lateral sclerosis [41]. CyPG, and in particular 15d-$PGJ_2$ have been detected also in experimental cellular and animal models of inflammation. The exploration of the biological effects of these eicosanoids has led to the observation of their antitumoral, antiviral and antiinflammatory effects. The antiviral effects of cyPG appear to be associated with the induction of a heat shock response in cells, the induction of IκBα expression and the inhibition of NF-κB activity [42, 43]. Recently, 15d-$PGJ_2$ has been identified as a potential endogenous ligand of the transcription factor known as PPARγ (peroxisome proliferator activated receptor γ) [44, 45]. The PPAR subfamily of nuclear receptors is constituted by three main forms, known as α, β and γ, which differ in agonist selectivity and tissue distribution. PPAR are ligand activated transcription factors involved in the modulation of lipid metabolism and inflammation [46, 47]. Since PPAR activation was found to counteract inflammation in many experimental models, this provided a mechanism for the antiinflammatory effects of cyPG. Based on the detection and the effects of 15d-$PGJ_2$ at the late stages of inflammation in a rat model of pleurisy, an antiinflammatory role for COX-2 was proposed [24]. During the last several years, negative effects on cyPG on proinflammatory gene expression have been frequently reported. CyPG have been shown to inhibit the induction of numerous genes involved in the pathophysiology of inflammation, including iNOS, COX-2, ICAM-1, MCP-1 and matrix metalloproteinase-9 (MMP-9) [48, 49]. For these reasons the potential for the pharmacological application of these or related agents as antiinflammatory drugs is being pursued.

A common mechanism underlying the antiinflammatory effects of cyPG in the various experimental systems is the inhibition of NF-κB. This effect can be due to several mechanisms. As stated above, some cyPG including 15d-$PGJ_2$ and $PGA_1$ can bind and activate PPAR. These transcription factors, in turn, may interfere with NF-κB activation by competing for limiting amounts of coactivators, by modulating the expression of IκB or by direct protein-protein interaction [50-52]. A PPAR-dependent effect of 15d-$PGJ_2$ has been reported to be responsible for the inhibition of iNOS and COX-2 expression in cardiomyocytes [53], and for the inhibition of iNOS and MMP-13 in human chondrocytes [54]. However, cyPG can also display antiinflammatory actions by PPAR-independent mechanisms. CyPG can induce oxidative stress and this in turn may be related to the induction of apoptosis in macrophages, a phenomenon that has been proposed to contribute to the resolution of inflammation [55], or mediate the induction of genes with antiinflammatory potential, such as heme-oxygenase [56].

The cyclopentenone moiety is an important structural feature of cyPG. This confers a strong reactivity to one of the carbons in the cyclopentenone ring (Figure 2) which may form adducts with cellular thiols, such as the sulphydryl group of glutathione or of protein cysteine residues [57], by a reaction known as Michael addition, which appears schematized in Figure 2B. Some cyPG possess a dienone structure with two electrophilic carbons (marked in Figure 2 by asterisks), both of which can be involved in the formation of Michael adducts [58, 59]. The reaction of cyPG with cysteine residues in proteins forming stable complexes can have important consequences for protein function [60]. The importance of this process for the antiinflammatory actions of cyPG is becoming more clear as the potential targets for this type of modification, which could be termed prostanylation or eicosanylation, are identified. We and others have described that cyPG may directly interfere with transcription factors involved in the control of inflammation, including NF-κB and AP-1 [59, 61, 62]. This effect may involve the interaction with proteins forming the signaling cascade that leads to transcription factor activation, or the interaction with transcription factors themselves. In this chapter we have used a fragment from the human p50 subunit of NF-κB, which corresponds to the DNA binding domain, as a model to explore the direct interaction of cyPG with transcription factors. Moreover, we have explored the effects of 15d-$PGJ_2$ on NF-κB signaling in a cellular model of inflammation, such as mesangial cells (MC).

# Materials and Methods

## Materials

15-Deoxy-$\Delta^{12,14}$-$PGJ_2$ was from Calbiochem-Novabiochem (San Diego, CA). 15-Deoxy-$\Delta^{12,14}$-$PGJ_2$ (mixture of isomers), $PGE_2$, $PGA_1$ and $PGA_1$-biotinimide were from Cayman Chemical (Ann Arbor, MI). Recombinant human IL-1β ($5 \times 10^7$ U/mg) was from Roche (Germany). Recombinant human TNF-α was from Serotec (Oxford, U.K.). Rabbit antisera against human p50 and p65 proteins were a generous gift of Dr. Nancy Rice (National Cancer Institute, Bethesda, MD, USA). Horseradish peroxidase (HRP) conjugated anti-rabbit immunoglobulins were from Dako (Glostrup, Denmark). Polyclonal anti-IκBα (sc-371) was from Santa Cruz Biotechnology (Santa Cruz, CA). The recombinant wild-type and mutant DNA binding domains of human p50 (amino acids 36 - 385, GenBank accession number M55643), expressed *in Escherichia coli* as hexahistidine fusion proteins and purified as described [34], were the generous gift of Dr. E. Pineda-Molina and Dr. S. Lamas. This construct will be referred to as p50. GSH was from Sigma. Ethylcarbodiimide, EZ-link 5-(biotinamido)pentylamine and biotinylated BSA standard were from Pierce (Rockford, IL). HRP-conjugated streptavidin was from Amersham Pharmacia Biotech (Buckinghamshire, U.K.). Cell culture media and supplements were from GIBCO Life Technologies S.A. (Barcelona, Spain).

## Preparation of Biotinylated 15d-$PGJ_2$

Biotinylated 15d-$PGJ_2$ was prepared by a procedure similar to that previously used for the preparation of biotinylated $PGA_2$ [63]. Briefly, 1 mg of 15d-$PGJ_2$ was made to react with 1 mg of EZ-link (5-biotinamido)pentylamine in the presence of the condensing agent diisopropylcarbodiimide. The reaction was carried out in acetonitrile for 16 h at room temperature. Biotinylated 15d-$PGJ_2$ was extracted with ethyl acetate and purified through a silica gel column eluted with ethyl acetate:methanol (9:1). The concentration of biotinylated PG was estimated from the determination of biotin as previously described [62]. The modified PG was then dried under argon and dissolved in DMSO for further use.

## Cells and Treatments

For this study, rat mesangial cells (RMC) between passages 10 and 15 were used. Primary cultures of RMC were obtained using a technique of graded sieving as reported earlier [15]. Groups of cells with mesangial morphology were subcloned and expanded. Mesangial cells displayed positive staining with anti-vimentin, anti-Thy 1.1, and anti-myosin, and absence of staining with anti-factor VIII [5]. Cells were grown in RPMI 1640 supplemented with 10% fetal bovine serum (FBS), 2 mM glutamine, 100 U/ml penicillin and 100 μg/ml streptomycin. For treatments, cells were incubated in RPMI medium without phenol red in the absence of serum. Cell activation was achieved by incubation with a combination of IL-1β (2 ng/ml) plus TNF-α (25 ng/ml) for the indicated time periods. Prostaglandins were added in DMSO to the incubation medium 15 min prior to cytokine addition. Final DMSO concentration was 0.1% (vol/vol).

## Analysis of NF-κB Transcriptional Activity

NF-κB transcriptional activity was estimated in transient transfection experiments. MC at 80% confluence were transfected with 1.5 μg of a luciferase reporter plasmid containing three consensus NF-κB response elements (3xNF-κB-pConA-Luc) [64] or empty vector (pConA-Luc), plus 2 ng of the Renilla luciferase reporter vector pRL-CMV (Promega, Madison, WI) for 3 h in Optimem (Gibco) using the LIPOFECTAMINE™ Reagent (Invitrogen-Life technologies, Barcelona, Spain). After a 24 h incubation in serum-free medium, cells were treated with 15d-$PGJ_2$ for 15 min before stimulation with cytokines, as indicated. Firefly and Renilla luminiscences were measured in cell lysates according to the instructions of the manufacturer using a dual luciferase reporter system from Promega. Promoter activities were expressed in arbitrary units as the ratio between the two luciferase measurements.

## Preparation of Nuclear and Cytosolic Extracts

Nuclear extracts from RMC were obtained as previously described [11]. RMC were resuspended in 10 mM Hepes, pH 7.9, 10 mM KCl, 0.1 mM EDTA, 0.1 mM EGTA, 1 mM dithiotreitol (DTT), 0.1 mM sodium vanadate, 50 mM sodium fluoride, 0.5 mM PMSF, and 2 μg/ml of leupeptin, antipain, trypsin inhibitor and pepstatin A. NP-40 was added to a final concentration of 0.5%. After vortexing for 10 s, nuclei were sedimented by centrifugation for 5 min at 15000 rpm. The supernatants were collected and centrifuged again at 200000 x g for 30 min to obtain the cytosolic fraction. The nuclear pellets were extracted by shaking on ice for 30 min in 20 mM Hepes, pH 7.9, 400 mM NaCl, 1 mM EDTA, 1 mM EGTA, 1 mM DTT, 0.1 mM sodium vanadate, 50 mM sodium fluoride plus protease inhibitors as above. Nuclear debris was discarded by centrifugation at 15000 rpm for 5 min, and nuclear extracts were stored at –80°C. Protein content was determined using the BCA protein assay from Pierce (Rockford, IL) for cytosolic extracts, and the Bradford method for nuclear extracts.

## Electrophoretic Mobility Shift Assays (EMSAs)

Sense and antisense oligonucleotides corresponding to the upstream NF-κB site from the human COX-2 promoter (5′-GGAGAGGGGATTCCCTGCG-3′), located within – 452 to – 433 bases from the transcriptional start site in the reported sequence [65] were annealed and labeled by incubation with [$\alpha$-$^{32}$P] dCTP using the klenow fragment, and isolated by gel filtration on Sephadex G-50. For the binding reaction, 2 ng of the labeled oligonucleotide (approximately 20000 cpm) and 1 μg of poly(dI-dC) carrier was incubated with 3.5 μg of nuclear extract in binding buffer (4% glycerol, 1 mM $MgCl_2$, 0.5 mM EDTA, 0.5 mM DTT, 50 mM NaCl, 10 mM Tris-HCl, pH 7.5) for 20 min on ice. Protein-DNA complexes were separated by electrophoresis on a non-denaturing 6% polyacrylamide gel and visualized by autoradiography. To ensure the specificity of the complexes formed, competition experiments were carried out by performing the binding assays in the presence of a 200-fold excess of unlabeled NF-κB oligonucleotide or of an unrelated oligonucleotide corresponding to bases – *456* to – *432* from the transcriptional start site in the reported sequence of the bovine endothelial NOS gene [66].

## Preparation of Cell Lysates and Western Blot

For analysis of protein levels in total cell lysates, cells were lysed in 10 mM Tris pH 7.5, 0.1 mM EDTA, 0.1 mM EGTA, 0.1 mM β-mercaptoethanol, 0.5% SDS, and 2 μg/ml of leupeptin, antipain, trypsin inhibitor and pepstatin A, by forced passes through a narrow needle on ice. Protein concentration was determined by the BCA protein assay from Pierce (Rockford, IL). Fifteen μg of protein from each experimental condition were electrophoresed on 10% polyacrylamide gels and transferred to Immobilon-P membranes (Millipore, Bedford, MA). As a control for even protein loading and transfer, membranes were stained with Ponceau S. Blots were probed with primary antibodies anti-IκBα (at 1:1000 dilution) or antip65 and anti-p50 (at 1:2000 dilution) followed by HRP-conjugated secondary antibodies at 1:2000 dilution, and immunocomplexes were visualized using an enhanced chemiluminiscence (ECL) detection system from Amersham. The intensity of the bands was estimated by image scanning and corrected by the values obtained from the Coomassie staining of membranes or by the signal given by an antibody against a non-related protein.

### Analysis of the NF-κB Dimers Present in Nuclear Extracts by Western Blot

Six μg of nuclear extracts were incubated with 400 ng of non-labeled NF-κB oligonucleotide, and the complexes formed were separated by gel electrophoresis, as above. Gels were transferred to Hybond-PVDF membranes and blots were probed with anti-p50 and anti-p65 antibodies before detection with ECL.

### Biotinylated 15d-PGJ$_2$ Labeling of p50 in RMC

Pre-confluent RMC were incubated with 10 μM biotinylated 15d-PGJ$_2$ for 2 h. Cells were washed twice with PBS, harvested and lysed in 10 mM Tris-HCl pH 7.5, 0.1 mM EDTA, 0.1 mM EGTA, 1% NP-40, 0.1 mM β-mercaptoethanol plus protease inhibitors. Lysates were centrifuged for 10 min at 10000 x g to discard insoluble debris. Supernatants containing 200 μg of protein were incubated batch-wise with 50 μl of Soft-Link Avidin (Promega, Madison, WI) for 1 h at 4°C. Beads were washed 5 times with lysis buffer by centrifugation at 6000 rpm for 1 min. Proteins retained on the beads were eluted by boiling the beads in Laemmli sample buffer. The presence of p50, or other proteins of interest, in the eluate was assessed by Western blot.

### Binding of Biotinylated cyPG to p50 in Vitro

The formation of covalent adducts between p50 and biotinylated cyPG, the structure of which is shown in Figure 3, was monitored by means of a Western blot-based assay. p50 in 20 mM Tris pH 7.5, 50 mM NaCl, 5mM $MgCl_2$, 1 mM EDTA, 0.7 μM β-mercaptoethanol, 0.1 mM DTT, 5% glycerol, 0.01% Nonidet P-40, was incubated with prostaglandins or vehicle (DMSO) at r.t. At the indicated time points Laemmli sample buffer was added and samples were analyzed by SDS-PAGE on 10% polyacrylamide gels. Proteins on gels were transferred to immobilon P membranes (Millipore) following the instructions of the manufacturer by using a semi-dry transfer blot apparatus from Bio-Rad. After blocking for 1 h by incubation with 2% non-fat dried milk in TBS-T (10 mM Tris-HCl, pH 7.5, 500 mM NaCl, 0.05% Tween 20), membranes were incubated with HRP-conjugated streptavidin (1:1000 dilution in 1% BSA in TBS-T). Signal was developed with the ECL system from Amersham. Blots were stripped by brief incubation with 8 N guanidinium hydrochloride, pH 3.0, and incubated with anti-p50 antibodies at 1:2000 dilution. Under the conditions used in the assays the incorporation of biotin was linear with respect to time, protein amount and biotinylated cyPG concentration. Estimation of biotin incorporation was achieved by comparison of the signal with that given by a biotinylated BSA standard (Pierce). After 4 h incubation of 5 μM p50 in the presence of 10 μM biotinylated 15d-$PGJ_2$ at room temperature, approximately 0.1 nmoles of biotin were incorporated per nmole of protein.

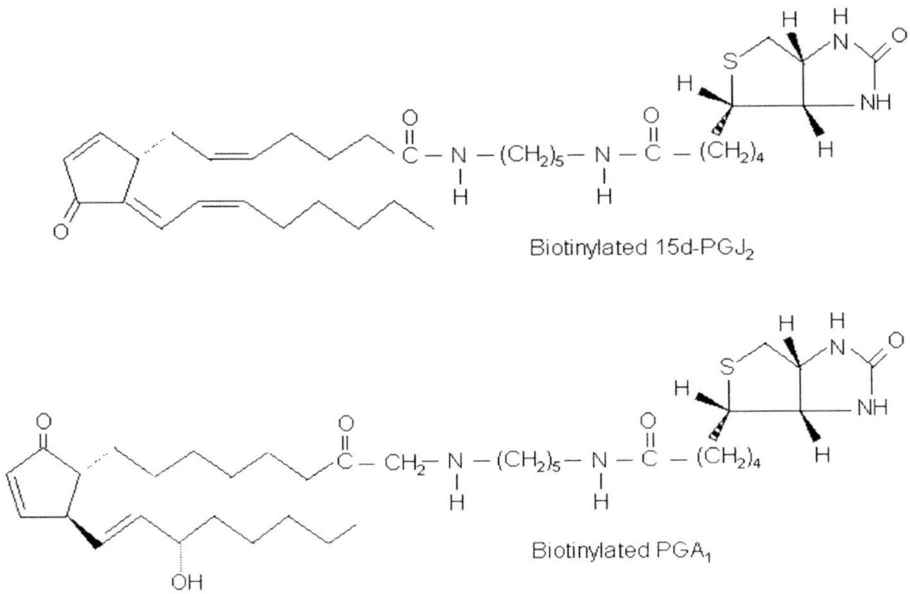


Figure 3. Structures of the biotinylated derivatives of 15d-$PGJ_2$ and $PGA_1$ used in this study.

# Results

## 15d-$PGJ_2$ Inhibits NF-κB Transcriptional Activity in RMC

Given the importance of the transcription factor NF-κB in the regulation of the induction of numerous proinflammatory genes, we were interested in studying the consequences of the interaction of cyPG with the NF-κB pathway in a model of inflammation such as RMC. For this purpose we first studied the effect of 15d-$PGJ_2$ on NF-κB transcriptional activity. RMC were transiently transfected with 1.5 μg of a NF-κB reporter plasmid containing 3 copies of NF-κB binding sequence driving the expression of luciferase (3xNF-κB-conA-luc) and 2 ng of pRL-CMV, as described in Methods. After a recovery period of 24 hours, RMC were treated with vehicle or 15d-$PGJ_2$ and 15 min later they were stimulated with IL-1β plus TNF-α (Ck) for 6 hours. The transcriptional activity of NF-κB was estimated from the luciferase activity present in cell lysates. Under these conditions the PG did not affect the basal activity of the reporter but inhibited cytokine-elicited induction of NF-κB activity by 30% (Figure 4).

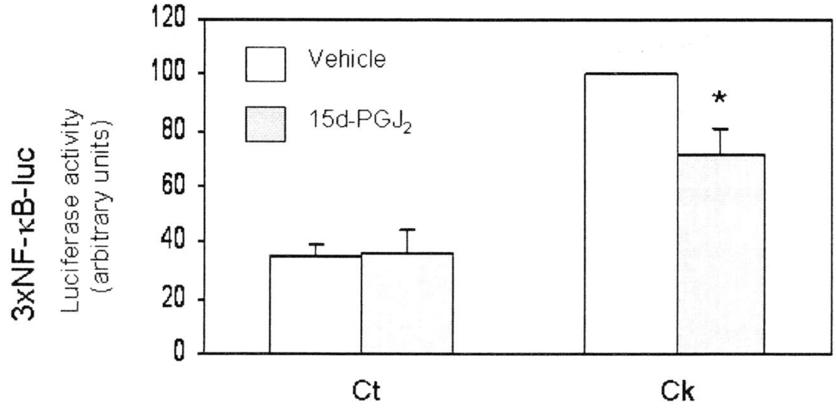


Figure 4. Effect of 15d-$PGJ_2$ on NF-κB transcriptional activity in RMC. The transcriptional activity of NF-κB in RMC was evaluated by transient transfection of a 3xNF-κB reporter construct. Cells were treated with 2 ng/ml IL-1β plus 25 ng/ml TNF-α (Ck) in presence or absence of 5 μM 15d-$PGJ_2$ for 6 hours. Results are average values ± SEM of three experiments performed in triplicate. (*$p<0.05$ *versus* vehicle).

## Effect of 15d-$PGJ_2$ on NF-κB DNA Binding Activity in Cytokine-stimulated RMC

We were then interested in exploring which steps in the NF-κB activation pathway were affected by 15d-$PGJ_2$ treatment. We first assessed whether 15d-$PGJ_2$ inhibited NF-κB DNA binding activity in intact cells. For this purpose we performed EMSA assays with nuclear extracts obtained from cells activated with Ck in presence or absence of the PG. Confluent RMC, kept in serum-free medium 24 hours before the experiment, were treated with the indicated agents at different time points. Activation of RMC with the cytokine mixture resulted in an induction in the formation of the NF-κB complex (Figure 5A), as previously described by us [11]. The specificity of this complex was assessed by competition assays performed in the presence of excess unlabeled NF-κB oligonucleotide (comp), which completely abolished binding of the radiolabeled probe, or of an unrelated oligonucleotide (unrelated), which did not block complex formation. In nuclear extracts from cells activated in the presence of 15d-$PGJ_2$ the NF-κB-DNA binding activity was clearly reduced. The inhibition of DNA-protein complex formation was apparent at all time points studied.

To gain insight into the mechanism of the inhibition of NF-κB-DNA binding we analyzed the levels of p50 and p65 subunits in the nuclear extracts by Western blot (Figure 5B). We observed that after 5 minutes of treatment with cytokines, there were not detectable differences in the nuclear levels of p65 and p50 in cells activated in the absence or presence of 15d-$PGJ_2$. However, at this time point, there was an intense inhibition of NF-κB-DNA complex formation (43%) in 15d-$PGJ_2$-treated cells.

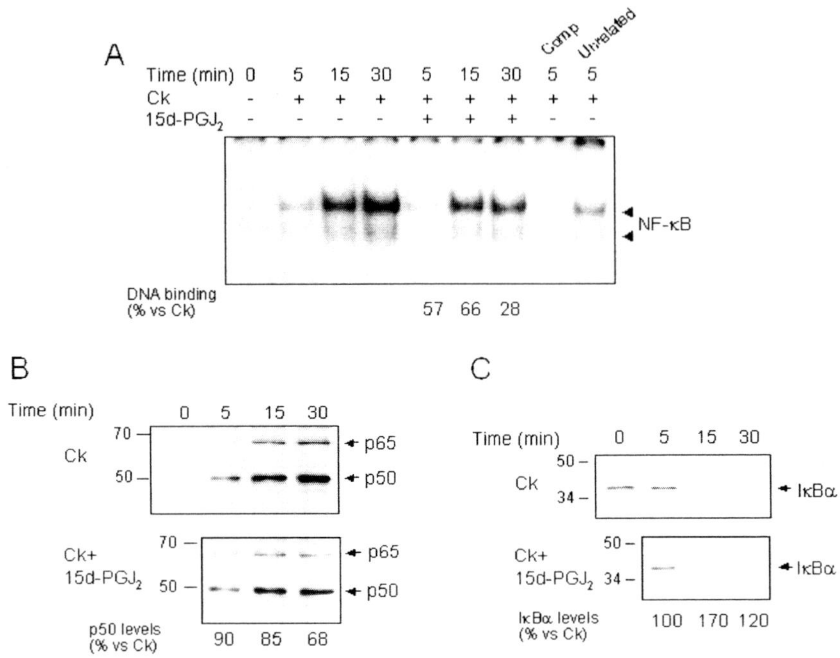


Figure 5. Effect of 15d-$PGJ_2$ on DNA binding and levels of NF-κB proteins. A) RMC were activated with Ck in presence or absence of 5 μM 15d-$PGJ_2$ and nuclear extracts were isolated at the indicated time points. The NF-κB-DNA binding activity was analyzed by EMSA. The formation of NF-κB-DNA complexes was quantitated and it is expressed as percentage of the values obtained in the absence of 15d-$PGJ_2$. Results shown are representative of two experiments. Levels of nuclear p50 and p65 (B) and cytosolic IκBα (C) were estimated by Western blot. The position of molecular mass standards (KDa) is shown on the left of the panels. Results are representative of three independent experiments.

This suggests that 15d-$PGJ_2$ may directly impair DNA binding in this model system. However, after 30 minutes of stimulation, there was a moderate reduction in the nuclear levels of p65 and p50 in 15d-$PGJ_2$-treated cells (32 % of the amount detected in cells treated with cytokines alone at the same time point). This suggests that a partial inhibition of NF-κB translocation to the nucleus may also contribute to the inhibitory actions of 15d-$PGJ_2$ on the NF-κB pathway. This possibility is also supported by the observation of an attenuation in the cytokine-elicited decrease in the cytosolic levels of IκBα in the cells that had been pre-treated with 15d-$PGJ_2$. After 15 min of cytokine stimulation, levels of IκBα were 1.7-fold higher in cells activated in the presence of 15d-$PGJ_2$ than in cells treated with cytokines alone (Figure 5C). These results suggest that both, an impairment of nuclear translocation of NF-κB subunits and a direct inhibition of NF-κB-DNA complex formation may contribute to the reduction in the NF-κB DNA binding activity elicited by 15d-$PGJ_2$ in RMC.

### 15d-PGJ$_2$ Inhibits NF-κB DNA Binding Activity in Vitro, without Disrupting NF-κB Dimers

To evidence a potential direct effect of cyPG on NF-κB-DNA complex formation we analyzed the effect of 15d-PGJ$_2$ on NF-κB DNA binding activity *in vitro* by EMSA. Confluent RMC were activated with Ck (IL-1β/TNFα) in serum-free medium during 15 minutes. Four micrograms of the nuclear extracts isolated as described in Methods, were incubated with different concentrations of prostaglandin before the binding assay. As observed in Figure 6A, treatment of nuclear extracts with 15d-PGJ$_2$ *in vitro* effectively inhibited NF-κB-DNA complex formation (43% inhibition in nuclear extracts treated with 10 μM 15d-PGJ$_2$ *versus* control).

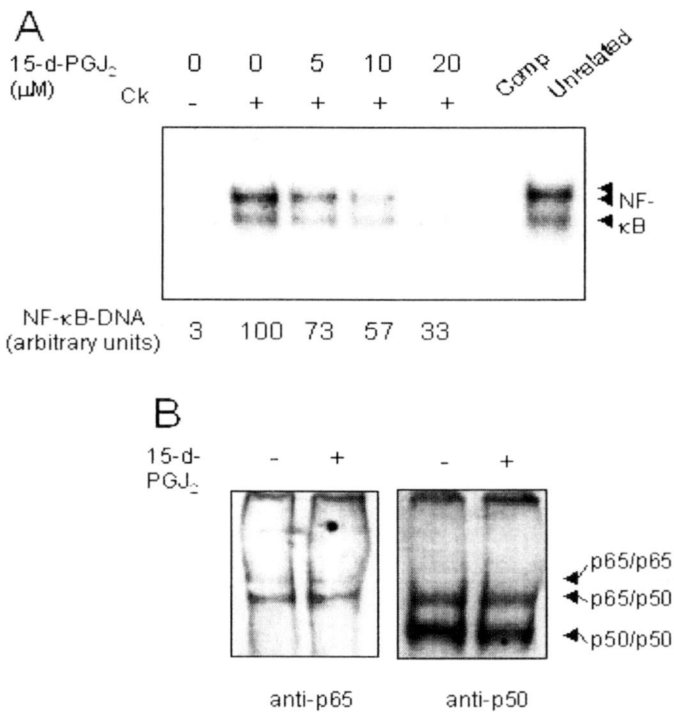


Figure 6. Effect of 15d-PGJ$_2$ on NF-κB DNA binding *in vitro*. A) RMC were activated with Ck for 15 min. Nuclear extracts were isolated and incubated with the indicated concentrations of 15d-PGJ$_2$ for 1 hour at r.t. The NF-κB-DNA binding activity was analyzed by EMSA. Competition assays were carried out using an excess of unlabeled NF-κB or unrelated oligonucleotide (200-fold). Protein-DNA complexes were quantitated by image scanning and the mean values from three experiments are shown. B) RMC were activated as before and the nuclear extracts were incubated in presence or absence of 5 μM 15d-PGJ$_2$ as above. The integrity of the dimers was analyzed by electrophoresis under non-denaturing conditions and Western blot with specific antibodies against p65 and p50. Results shown are representative of three experiments.

To discard the possibility that the inhibitory effect could be due to a disruption or degradation of the NF-κB dimers present in the nuclear extracts during the incubation with the PG, we performed an immunoblot of the non-denaturing gels run under the same conditions used for EMSA with specific antibodies against p50 and p65 subunits (Figure 6B). This Western blot showed the presence of protein bands which were detected only by anti-p65 antibodies, corresponding probably to p65 homodimers. A faster migrating band reacting with both anti-p65 and anti-p50 antibodies was also detected which was probably formed by p65/p50 heterodimers. Finally, the lower band, reacting only with anti-p50 antibodies was probably due to p50 homodimers. The p65/p50 and the p50 bands co-migrated with the main specific bandshifts observed in EMSAs performed with the labeled oligonucleotide. As it can be observed in Figure 6B, incubation of the isolated nuclear extracts in the presence of 15d-$PGJ_2$, in the conditions under which NF-κB-DNA binding was clearly inhibited, did not affect the integrity of the NF-κB dimers. These results indicate that the effect of 15d-$PGJ_2$ is probably due to an impairment in the formation of NF-κB-DNA complexes.

## Biotinylated 15d-$PGJ_2$ Binds to p50 and p65 Proteins in Intact RMC

We have previously shown that 15d-$PGJ_2$ may interact with p50 *in vitro* [62]. To explore whether 15d-$PGJ_2$ may directly interact with NF-κB proteins in RMC we incubated intact cells in the presence of a biotinylated 15d-$PGJ_2$ derivative (Figure 7). This compound retains the cyclopentenone moiety of the PG and it is able to modify cysteine residues in proteins by Michael addition (see Figure 3).

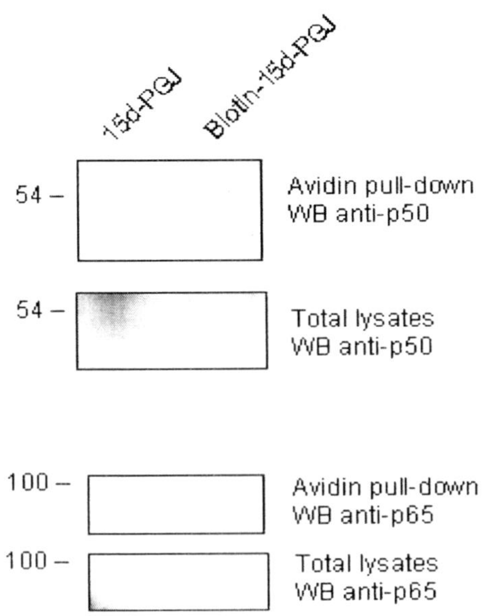


Figure 7. Biotinylated 15d-$PGJ_2$ binds to NF-κB proteins in RMC. RMC were incubated with 10 μM 15d-$PGJ_2$ or biotinylated 15d-$PGJ_2$ in serum-free medium for 2 h. Cell lysates were incubated with avidin beads. After extensive washing, proteins retained on the beads were eluted in Laemmli buffer and analyzed by gel electrophoresis and Western blot with antibodies against the p50 and p65 subunits of NF-κB, as indicated. The levels of NF-κB subunits in total cell lysates are shown. Results are representative of two experiments.

After incubation with biotinylated 15d-$PGJ_2$, proteins with biotin moieties were collected by incubation with avidin-agarose beads and the proteins of interest were detected in the eluate by Western blot. This type of assay showed the presence of both p50 and p65 NF-κB subunits in the avidin-binding fraction of RMC incubated with biotinylated 15d-$PGJ_2$ but not in the control samples from RMC incubated with non-biotinylated PG. These results suggest that biotinylated 15d-$PGJ_2$ is able to form covalent adducts with NF-κB proteins in RMC.

## Incorporation of Biotinylated cyPG into the NF-κB Subunit p50 in Vitro is Time and Dose-Dpendent

We were then interested in exploring the structural determinants involved in the interaction between NF-κB proteins and cyPG. For this purpose we have set up the conditions for a Western blot-based assay that would allow us to monitor the *in vitro* modification of p50 by biotinylated cyPG in a semi-quantitative manner. As it is shown in Figure 8, the incorporation of biotinylated 15d-$PGJ_2$ into the p50 construct was linear with time at least up to 4 h incubation.

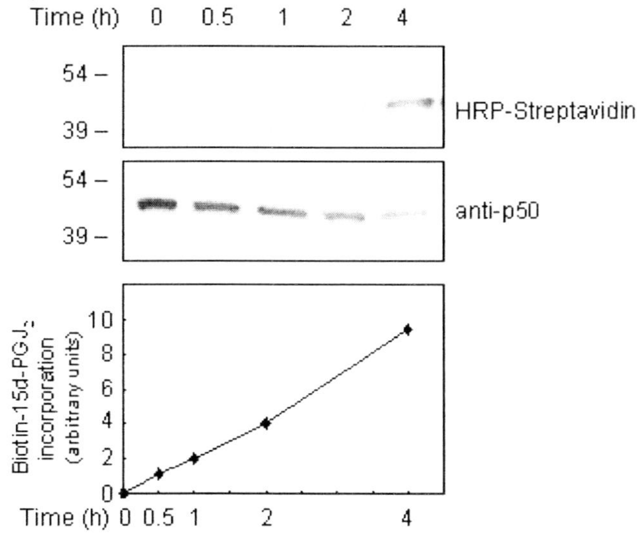


Figure 8. Time-course of the incorporation of biotinylated 15d-$PGJ_2$ into p50. Human recombinant p50 construct at 0.5 μM was incubated with 1 μM biotinylated 15d-$PGJ_2$ at r.t. for the indicated time periods. Incubation mixtures were analyzed by SDS-PAGE and Western blot and incorporation of biotin was detected with HRP-conjugated streptavidin and ECL. The amount of p50 present in the incubation was assessed by Western blot with anti-p50 antibody. The intensity of the biotin signal, normalized by the signal given by the anti-p50 antibody, and expressed in arbitrary units, is shown in the lower panel. Results are representative of three assays.

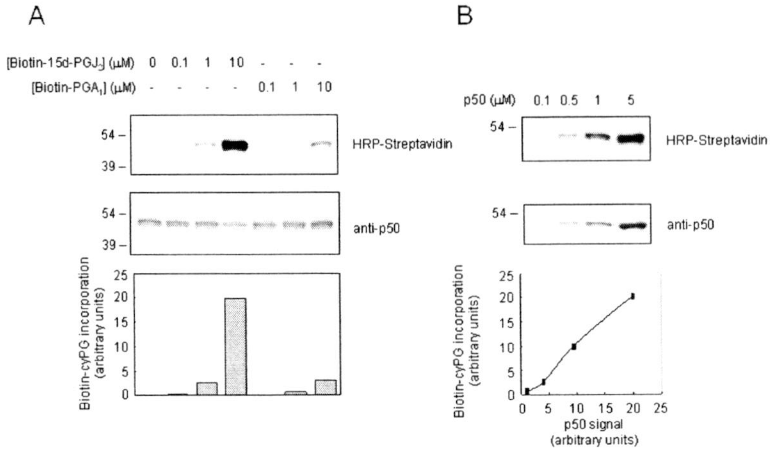


Figure 9. Dose-response of the incorporation of biotinylated cyPG into p50. (A) Recombinant p50 at 0.5 µM was incubated in the presence of the indicated concentrations of biotinylated cyPG for 1 h at r.t. Biotinylated cyPG incorporation was estimated as described in the legend of Figure 8. (B) The indicated concentrations of p50 were incubated with 1 µM biotinylated 15d-PGJ2 at r.t. for 1 h. The lower panel shows the plot of the biotin signal versus the p50 signal. Results are representative of two assays.

Figure 9 shows the concentration-dependence of this assay. Incorporation of biotinylated cyPG was dependent on the concentration of the PG (Figure 9A).Interestingly, under the same conditions, the modification of p50 by biotinylated 15d-PGJ$_2$ was more intense than with biotinylated PGA$_1$. This difference may be due to the different reactivity of the two PG or to factors related to the structural features of the protein. As it is shown in Figure 9B, 15d-PGJ$_2$ binding to p50 in this assay was also linear with respect to the concentration of the p50 construct. This indicates that this type of assay provides a semi-quantitative approach to study the factors which influence the binding of biotinylated cyPG to p50 *in vitro*.

## Binding of Biotinylated 15d-PGJ$_2$ to p50 is Selective

In a previous study we have reported that *in vitro* treatment of p50 with 15d-PGJ$_2$ strongly diminishes its DNA binding ability [62]. PGA$_1$ exerted a moderate inhibitory effect, while carbaprostacyclin and PGE$_2$, which are not cyclopentenones, did not inhibit the binding. Here we have used the Western blot assay to explore the ability of several PG to block the binding of biotinylated 15d-PGJ$_2$ to p50. The results depicted in Figure 10 show that the binding of biotinylated 15d-PGJ$_2$ to p50 can be blocked by an excess of 15d-PGJ$_2$. Excess PGA$_1$ partially reduced labeling while PGE had no effect. These observations suggest that the binding of 15d-PGJ$_2$ to p50 is selective and that the cyclopentenone moiety is an important determinant for the interaction.

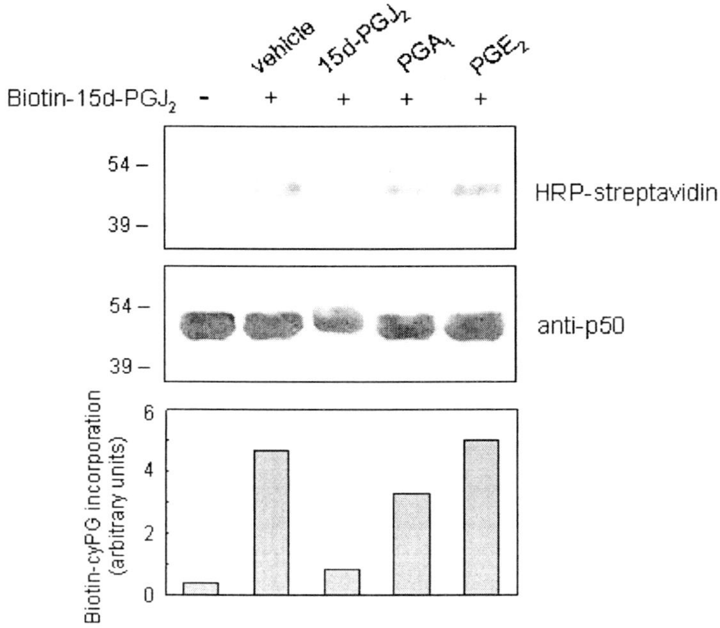


Figure 10. Effect of several PG on the binding of biotinylated 15d-$PGJ_2$ to p50. p50 at 0.5 μM was incubated with 1 μM biotinylated 15d-$PGJ_2$ in the absence or presence of the indicated PG at 100 μM. Biotin incorporation was assessed as above. Results are representative of two assays.

## Biotinylated cyPG Bind to Cysteine 62 in Human p50, in a Manner Dependent on the Interaction with DNA

The ability of cyPG to inhibit NF-κB-DNA binding suggests that binding of cyPG directly precludes the interaction of p50 with DNA. We then explored whether p50 binding to DNA could affect the incorporation of biotinylated 15d-$PGJ_2$. As it is shown in Fig. 11A, incubation of p50 with a double stranded oligonucleotide bearing the sequence of one of the NF-κB binding sites from the human COX-2 promoter, as detailed in Methods, inhibited the binding of biotinylated 15d-$PGJ_2$ by approximately 70%. In contrast, an NF-κB oligonucleotide with a mutated sequence elicited only a 40% inhibition, while a nucleotide with a consensus sequence for the binding of AP-1 was virtually without effect. Therefore, the interaction of p50 with DNA exerts a protective effect on the modification of critical cysteine residues by cyPG. This suggests that the interaction of 15d-$PGJ_2$ with p50 involves residues closely implicated in DNA binding. Evidence obtained in previous work strongly suggested that cysteine 62 of human p50, which is located in the DNA binding domain of this protein, was involved in the interaction with 15d-$PGJ_2$ [62, 67]. Here we have explored whether this cysteine residue plays also a role in the interaction of p50 with biotinylated $PGA_1$. As it is shown in Figure 11B, the incorporation of both biotinylated 15d-$PGJ_2$ and $PGA_1$ into p50 was strongly reduced (more than 90% inhibition) when cysteine 62 was mutated to serine. These results indicate that cysteine 62 is an important target for the binding of both A-type and J-type cyPG to p50.

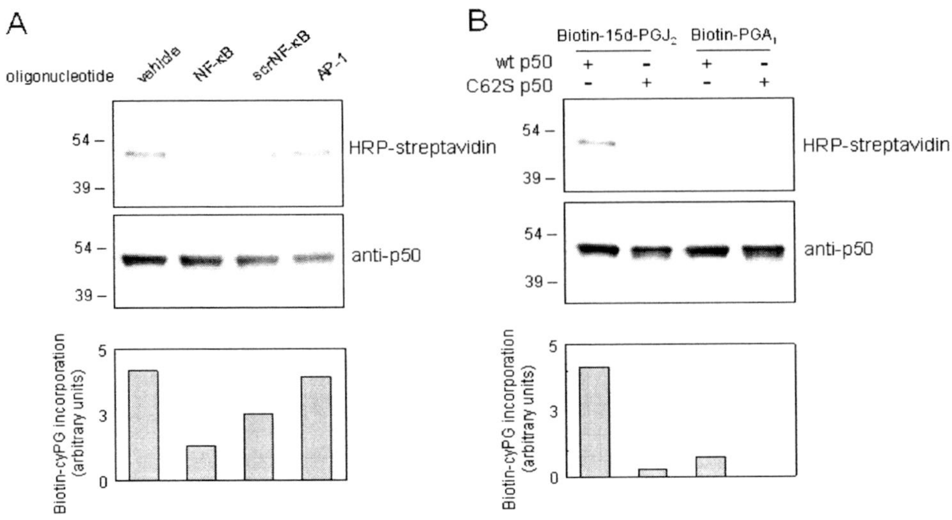


Figure 11. Involvement of the DNA binding region of p50 in the interaction with biotinylated cyPG. (A) Effect of different oligonucleotides on the binding of biotinylated 15d-$PGJ_2$ to p50. p50 at 0.5 μM was incubated in the presence of the indicated oligonucleotides or annealing buffer (vehicle) for 15 min at r.t., after which, biotinylated 15d-$PGJ_2$ at 1 μM final concentration was added and the incubation was continued for 1 h. NF-κB, 5'-GGAGAGGGGATTCCCTGCG-3'; scrNF-κB, 5'- GGACGAGCTGCGT GCGTGA-3'; AP-1, 5'-TGTTGGTGACTAATAACAC-3'. (B) Wild type (wt) or cysteine 62 to serine mutant p50 (C62S) at 0.5 μM was incubated with 1 μM biotinylated 15d-$PGJ_2$ or $PGA_1$. Biotin incorporation and p50 levels were assessed as described in the legend of Figure 8.

## CyPG Interact with AP-1 Proteins

Evidence available indicate that the cellular targets of cyPG are multiple. We were then interested in exploring the presence of other potential targets for modification by cyPG in RMC. Incubation of intact RMC with biotinylated 15d-$PGJ_2$ resulted in the incorporation of the biotin label into multiple bands, as detected by gel electrophoresis and Western blot with HRP-conjugated streptavidin (Figure 12A). The bands detected were clearly different from the endogenous biotinylated proteins. This indicates that the modification of several potential protein targets may contribute to the effects of cyPG in RMC. Among these targets we have recently reported that the component of AP-1 c-Jun can be modified by biotinylated 15d-$PGJ_2$ in murine fibroblasts [59]. 15d-$PGJ_2$ has been shown to inhibit the activity of AP-1 in several experimental systems [48, 68], and this effect could contribute to the antiinflammatory actions of cyPG. The potential modification of AP-1 by biotin-15d-$PGJ_2$ in mesangial cells was explored by means of an avidin pull-down assay. As it is shown in Figure 12B, c-Jun present in RMC lysates was retained on avidin beads when cells had been incubated with biotinylated 15d-$PGJ_2$ but not when cells had been treated with non biotinylated PG. This indicates that in this cell type, AP-1 components may also be targets for cyPG.

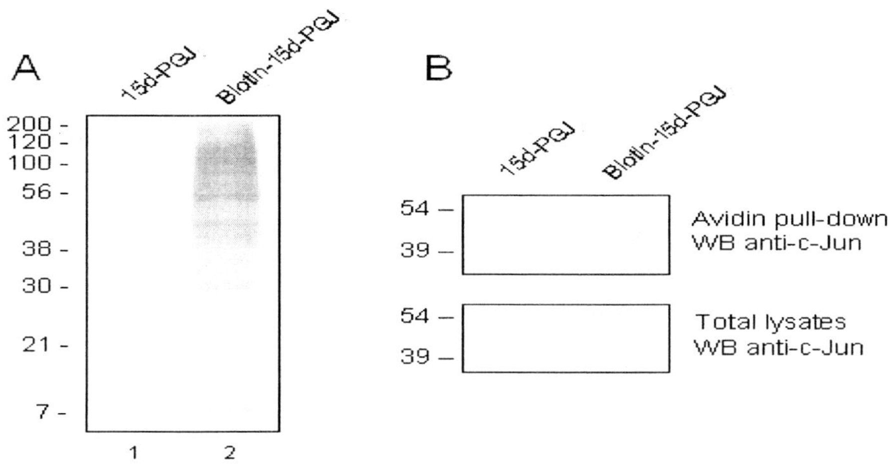


Figure 12. The transcription factor AP-1 is a target for biotinylated 15d-$PGJ_2$ binding in RMC. (A) RMC were incubated for 2 h with 10 μM 15d-$PGJ_2$ or biotinylated 15d-$PGJ_2$. Aliquots from cell lysates containing 10 μg of protein were analyzed by SDS-PAGE, Western blot and detection with HRP-streptavidin. Biotin positive polypeptides were visualized by ECL. Bands detected in lane 1 correspond to endogenous biotinylated proteins. (B) Cell lysates were incubated with avidin beads. After extensive washing, proteins retained on the beads were eluted with Laemmli sample buffer. The presence of c-Jun in the eluates was assessed by Western blot and detection with HRP-conjugated streptavidin and ECL. Results shown are representative of three assays.

# Discussion

Evidence accumulated during the last several years indicate that cyPG at micromolar concentrations exert potent antiinflammatory effects both in cellular systems and in animal models of inflammation. The mechanisms responsible for these effects appear to be multiple and they have been only partially characterized. Inhibition of signal transduction pathways by cyPG and direct interference with transcription factors have been reported. Given their importance for the inflammatory response, the transcription factors NF-κB and AP-1 have been studied as potential targets for the antiinflammatory actions of cyPG. Inhibitory effects of cyPG on NF-κB activity have been previously described, and they have been shown to correlate with the antiinflammatory potency of cyPG *in vivo*. Inhibition of the expression of proinflammatory genes in association with NF-κB blockade has been proposed to underlie the beneficial effects of cyPG treatment on acute renal failure, restenosis after balloon angioplasty in rats or carrageenin-induced pleurisy [69-71]. In this chapter we have described that the cyPG 15d-$PGJ_2$ reduces the transcriptional activity of NF-κB elicited by cytokines in RMC, as assessed by transfection of a reporter plasmid. CyPG and in particular 15d-$PGJ_2$ may interfere with the NF-κB activity at multiple levels. The two main mechanisms characterized to date are related to the ability of cyPG to activate PPAR or to modify cellular proteins by Michael addition. Several proteins along the NF-κB activation pathway have been proposed to be targets for modification by cyPG. CyPG has been reported to directly impair the function of activated IKK both in intact cells and *in vitro*, reducing its ability to phosphorylate IκB which results in decreased degradation of IκB and impaired nuclear translocation of NF-κB dimers [72, 73]. The requirement of a cysteine residue at position 179 in the activation loop of IKKβ for this inhibition to occur, suggested the interaction between the cyclopentenone moiety of the cyPG and the sulphydryl group of this residue by Michael addition. More recently, adducts of biotinylated cyPG and both, IKKα and β have been reported [74, 75]. The contribution of this interaction to the inhibition of NF-κB appears to depend on the cell system under study. In HeLa cells, inhibition of NF-κB transcriptional activity and DNA binding does not involve the inhibition of IKK activity, IκB degradation or nuclear translocation of NF-κB subunits [61, 62]. In contrast, in murine macrophages cyPG impair IκBα phosphorylation and degradation, thus pointing to the involvement of the inhibition of IKK [61, 72]. In addition cyPG can interfere with the activity of the proteasome [76], and a biotinylated 15d-$PGJ_2$-proteasome conjugate has been recently identified [77]. The results presented in this chapter suggest that in RMC cyPG may impair IκBα degradation, since the reduction in IκBα levels elicited by cytokines is attenuated in cells activated in the presence of 15d-$PGJ_2$. However, whether this effect is due to an inhibition of IκB phosphorylation by IKK or to decreased proteasome activity requires further investigation. In addition, our results show that 15d-$PGJ_2$ can inhibit the DNA binding activity of NF-κB more intensely than the nuclear translocation of NF-κB subunits p65 and p50, and that this inhibition can also occur in isolated nuclear extracts. These results suggest that 15d-$PGJ_2$ can directly interfere with NF-κB DNA binding in RMC. These results are in agreement with previous reports in which cyclopentenones have been shown to block the binding of recombinant p65 [61] or p50 [62] to DNA. The requirement for the presence of specific cysteine residues, cysteine 38 in human p65 and cysteine 62 in human p50, for this inhibition to occur, points to the involvement of a direct interaction between the cyclopentenone moiety of the PG and the sulphydryl groups in proteins. In this chapter we provide evidence for the formation of covalent adducts between biotinylated 15d-$PGJ_2$ and p65 and p50 in intact RMC. These results confirm previous observations from our group in HeLa cells [62]. A schematic view of the NF-κB activation pathway with the potential sites of interaction of cyPG is shown in Figure 13. In addition, we have monitored this interaction by a semi-quantitative assay using recombinant p50 *in vitro*. Under the conditions of our assay the modification of p50 by 15d-$PGJ_2$ can be detected with concentrations of the PG as low as 100 nM, and it is more efficient than the modification by $PGA_1$. At present it is not known whether this differential effect might be due to structural determinants of the PG or the protein. This will be the subject of future investigations. The modification of p50 by cyPG *in vitro* appears to involve a region which is masked by DNA binding, since incubation in the presence of an oligonucleotide with the consensus sequence for NF-κB binding reduces incorporation of the PG. The cysteine residue located at position 62 in the sequence of human p50 has been shown to be important for DNA binding. Modification of this residue by oxidation or alkylating agents results in loss of DNA binding activity [34]. Substitution of this cysteine residue by serine renders the p50 protein virtually insensitive to redox-induced inhibition of DNA binding [34]. Interestingly, under our experimental conditions, the incorporation of biotinylated cyPG into the cys62ser mutant is reduced by more than 90%. This indicates that this cysteine residue is the main site of attachment for both 15d-$PGJ_2$ and $PGA_1$ derivatives *in vitro*.

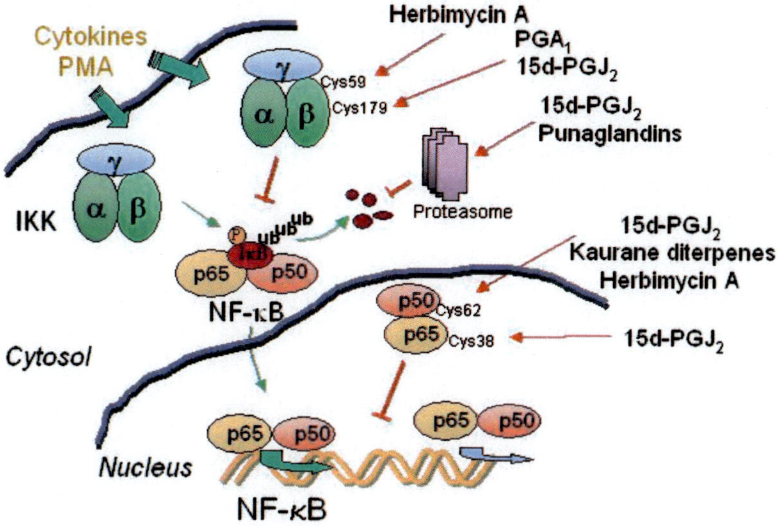


Figure 13. Schematic view of the NF-κB activation pathway showing the sites of interaction of several Michael receptors. Given the importance of the transcription factor NF-κB for the inflammatory response, the pathway leading to its activation has been the subject of numerous studies in search for antiinflammatory molecules. Shown is the site of action of several molecules which have been reported to interfere with NF-κB activation through the formation of covalent adducts with critical proteins of this pathway by Michael addition.

We and other authors have reported that cyPG can interact with the AP-1 pathway by PPAR-dependent and independent mechanisms [59, 60]. These interactions may result either in activation of inhibition of AP-1 depending on the cell model under study. Treatment of RMC with 15d-$PGJ_2$ has been reported to result in a reduction in AP-1 DNA binding [59, 68]. Our results show that in RMC the component of the AP-1 transcription factor c-Jun may also be a target for modification by biotinylated 15d-$PGJ_2$. This suggests that inhibition of AP-1 by direct modification with cyPG could contribute to the antiinflammatory effects of cyPG in RMC.

Not only NF-κB/Rel proteins and c-Jun, but other transcription factors involved in the inflammatory response possess critical cysteine residues susceptible to redox regulation that could be targeted by direct modification with cyPG. Examples of these factors are cAMP response element binding protein (CREB), glucocorticoid receptor and hypoxia inducible factor 1α (HIF-1α), among others. Therefore, the possibility exists that cyPG can form covalent adducts with other transcription factors not yet identified. In addition, as detailed above for NF-κB, the function of transcription factors important for the cellular defense against stress or for inflammation can be regulated by the direct modification not of the transcription factors themselves but of cytosolic or nuclear proteins involved in the regulation of their activity. This is the case of the protein Keap1, which controls the subcellular localization and degradation of the transcription factor Nrf2, involved in the regulation of genes encoding antioxidant enzymes. Modification of Keap1 by cyPG leads to activation of Nrf2 [78, 79]. This type of indirect mechanism could also influence the activity of the factors subject of this study, since both thioredoxin and thioredoxin reductase, proteins involved in the maintenance of the reduced state of critical cysteine residues in AP-1, Rel and p53 proteins, have been shown to be modified and inhibited by cyPG [75, 80].

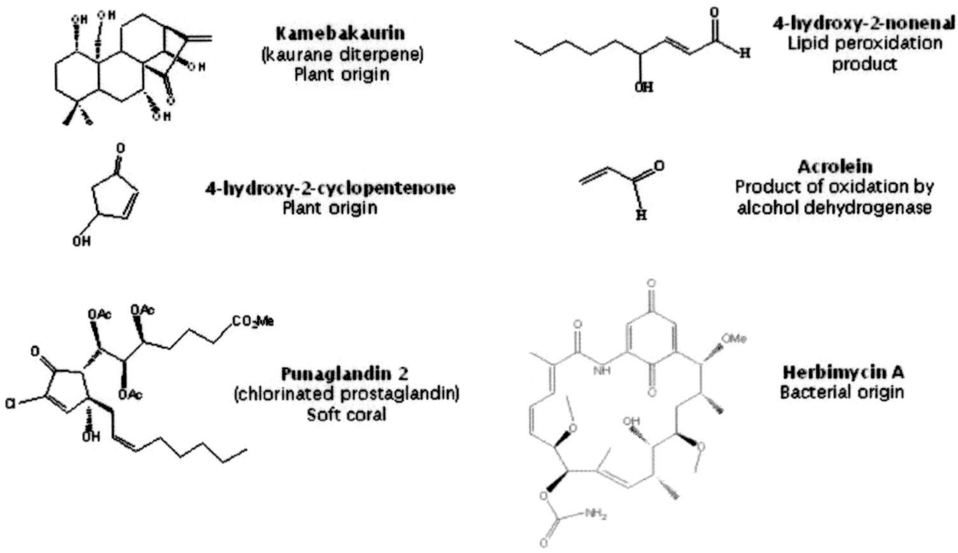


Figure 14. Structure of Michael receptors of diverse origin.

In the light of the importance of covalent protein modification in the antiinflammatory effects of cyPG, it is interesting to consider the potential of other compounds with unsaturated carbonyl groups that could act by similar mechanisms. This is the case of several compounds of plant origin, the structure of some of which is shown in Figure 14. The kaurane diterpene kamebakaurin is present in plants used in oriental medicine for the treatment of cancer and inflammation. The mechanism of action of this compound has been reported to involve the formation a covalent adduct with the p50 subunit of NF-κB, targeting cysteine 62 of p50 in a manner analogous to 15d-$PGJ_2$ [81]. The compound 4-hydroxy-2-cyclopentenone is present in the extracts of *Passiflora tetranda* [82], and it is responsible for the antibacterial properties of this preparation which has been used in herbal medicine. Recently, a class of compounds termed punaglandins has been isolated from the soft coral *Telesto riisei* [83], and they have been reported to exhibit potent antiinflammatory and antiproliferative effects. The molecular basis for these actions may rely, at least in part, in their ability to function as potent Michael receptors to inhibit the ubiquitin isopeptidase activity of the proteasome. In addition, several products of lipid peroxidation such as 4-hydroxy-2-nonenal or acrolein possess unsaturated carbonyl groups that may form covalent adducts with proteins. Proteins carrying these oxidative modifications may be more resistant to degradation and accumulate in cells, thus contributing to the pathophysiology of various disease processes including atherosclerosis, diabetes or Alzheimer's disease [84, 85]. Unsaturated carbonyl groups are also present in certain drugs and enzyme inhibitors, like ethacrynic acid or herbimycin. The tyrosine kinase inhibitor herbimycin A has been found to block NF-κB activation by IL-1 or phorbol 12-myristate 13-acetate (PMA) in several cell types through the interaction with cysteine 62 in p50 [86]. Recently, the interaction of herbimycin with the β subunit of IKK has been reported [87]. This interaction, which abrogates IKK activity and NF-κB activation has been reported to occur at cysteine 59, which is located near the catalytic site of IKK. Therefore, in a manner analogous to cyPG, herbimycin A can also target several steps in the NF-κB activation pathway and for this reason it has been proposed as a novel candidate for antiinflammatory drug. The sites of action of some Michael receptors along the NF-κB activation pathway is shown in Figure 13.

In summary, cyPG represent a class of compounds with high reactivity towards cellular thiols which exert varied biological actions of potential pharmacological interest. Among the targets of cyPG, the NF-κB and AP-1 activation pathways constitute important points of intervention for the control of inflammatory processes. Elucidation of the mechanisms underlying the effects of cyPG and identification of other potential targets will contribute to assess their potential as antiinflammatory or antiviral molecules.

## Acknowledgements

We wish to acknowledge the work of collaborators who have contributed to the research on this topic, Francisco J. Sánchez-Gómez, Konstatinos Stamatakis, Dr. E. Pineda-Molina and Dr. F. J. Cañada, and the technical assistance of M. J. Carrasco. Our work is supported by Grants SAF2003-03713 from Ministerio de Educación y Ciencia and 08.4/0025.1/2003 from Comunidad de Madrid (Spain).

## References

[1] Nathan, C. (2002) Points control in inflammation. *Nature*, *420*, 846-852.

[2] Kunz, D., Walker, G., Eberhardt, W. & Pfeilschifter, J. (1996) Molecular mechanisms of dexamethasone inhibition of nitric oxide synthase expression in interleukin 1β-stimulated mesangial cells: evidence for the involvement of transcriptional and posttranscriptional regulation. *Proc. Natl. Acad. Sci. USA*, *93*, 255-259.

[3] Laskin, J. D., Heck, D. E. & Laskin, D. L. (1994) Multifunctional role of nitric oxide in inflammation. *Trends Endocrinol. Metab.*, *5*, 377-382.

[4] Pérez-Sala, D. & Lamas, S. (2001) Regulation of cyclooxygenase-2 expression by nitric oxide in cells. *Antioxid. Redox Signal.*, *3*, 231-248.

[5] Saura, M., Martínez-Dalmau, R., Minty, A., Pérez-Sala, D. & Lamas, S. (1996) Interleukin-13 inhibits inducible nitric oxide synthase expression in human mesangial cells. *Biochem. J.*, *313*, 641-646.

[6] Pérez-Sala, D., Díaz-Cazorla, M., Ros, J., Jiménez, W. & Lamas, S. (1997) Tetrahydrobiopterin modulates cyclooxygenase-2 expression in human mesangial cells. *Biochem. Biophys. Res. Commun.*, *241*, 7-12.

[7] Saura, M., Zaragoza, C., Díaz-Cazorla, M., Hernández-Perera, O., Eng, E., Lowenstein, C. J., Pérez-Sala, D. & Lamas, S. (1998) Involvement of transcriptional

mechanisms in the inhibition of NOS2 expression by dexametnasone in rat mesangial cells. *Kidney Int.*, *53*, 38-49.

[8] Tetsuka, T., Daphna-Iken, D., Srivastava, S. K., Baier, L. D., DuMaine, J. & Morrison, A. R. (1994) Cross-talk between cyclooxygenase and nitric oxide pathways: prostaglandin $E_2$ negatively modulates induction of nitric oxide synthase by interleukin 1. *Proc. Natl. Acad. Sci. USA*, *91*, 12168-12172.

[9] Mühl, H. & Pfeilschifter, J. (1994) Tetrahydrobiopterin is a limiting factor of nitric oxide generation in interleukin 1ß-stimulated rat glomerular mesangial cells. *Kidney Int.*, *46*, 1302-1306.

[10] Tetsuka, T., Daphna-Iken, D., Miller, B. W., Guan, Z., Baier, L. D. & Morrison, A. R. (1996) Nitric oxide amplifies interleukin 1-induced cyclooxygenase-2 expression in rat mesangial cells. *J. Clin. Invest.*, *97*, 2051-2056.

[11] Díaz-Cazorla, M., Pérez-Sala, D. & Lamas, S. (1999) Dual effect of nitric oxide donors on cyclooxygenase-2 expression in human mesangial cells. *J. Am. Soc. Nephrol.*, *10*, 943-952.

[12] Schlöndorff, D. (1996) Roles of the mesangium in glomerular function. *Kidney Int.*, *49*, 1583-1585.

[13] Mené, P. (1996) Physiology and pathophysiology of the mesangial cell. *Nefrología*, *16*, 8-13.

[14] Rzymkiewicz, D., Leingang, K., Baird, N. & Morrison, A. (1994) Regulation of prostaglandin endoperoxide synthase gene expression in rat mesangial cells by interleukin-1 beta. *Am. J. Physiol.*, *266*, F39-F45.

[15] Saura, M., López, S., Rodríguez-Puyol, M., Rodríguez-Puyol, D. & Lamas, S. (1995) Regulation of inducible nitric oxide synthase expression in rat mesangial cells and isolated glomeruli. *Kidney Int.*, *47*, 500-509.

[16] Salvemini, D., Settle, S. L., Masferrer, J. L., Seibert, K., Currie, M. G. & Needleman, P. (1995) Regulation of prostaglandin production by nitric oxide; an in vivo analysis. *Br. J. Pharmacol.*, *114*, 1171-1178.

[17] Salvemini, D. & Masferrer, J. L. (1996) Interactions of nitric oxide with cyclooxygenase: in vitro, ex vivo and in vivo studies. *Methods Enzymol.*, *269*, 12-25.

[18] Nogawa, S., Forster, C., Zhang, F., Nagayama, M., Ross, M. E. & Iadecola, C. (1998) Interaction between inducible nitric oxide synthase and cyclooxygenase- 2 after cerebral ischemia. *Proc. Natl. Acad. Sci. U S A*, *95*, 10966-10971.

[19] von Knethen, A. & Brüne, B. (1997) Cyclooxygenase-2: an essential regulator of NO-mediated apoptosis. *FASEB J.*, *11*, 887-895.

[20] von Knethen, A. & Brune, B. (2000) Superinduction of cyclooxygenase-2 by NO(*) and agonist challenge involves transcriptional regulation mediated by AP-1 activation. *Biochemistry*, *39*, 1532-1540.

[21] Habib, A., Bernard, C., Lebret, M., Créminon, C., Esposito, B., Tedgui, A. & Maclouf, J. (1997) Regulation of the expression of cyclooxygenase-2 by nitric oxide in rat peritoneal macrophages. *J. Immunol.*, *158*, 3845-3851.

[22] Pérez-Sala, D., Cernuda-Morollón, E., Díaz-Cazorla, M., Rodríguez-Pascual, F. & Lamas, S. (2001) Posttranscriptional regulation of human iNOS by the NO/cGMP pathway. *Am. J. Physiol. Renal Physiol.*, *280*, F466-F473.

[23] Marotta, P., Sautebin, L. & Di Rosa, M. (1992) Modulation of the induction of nitric oxide synthase by eicosanoids in the murine macrophage cell line J774. *Br. J. Pharmacol.*, *107*, 640-641.

[24] Gilroy, D. W., Colville-Nash, P. R., Willis, D., Chivers, J., Paul-Clark, M. J. & Willoughby, D. A. (1999) Inducible cyclooxygenase may have anti-inflammatory properties. *Nat. Med.*, *5*, 698-701.

[25] Levy, B. D. (2001) Lipid mediator class switching during acute inflammation: signals in resolution. *Nat. Immunol.*, *2*, 612-619.

[26] Karin, M. (1995) The regulation of AP-1 activity by mitogen-activated protein kinases. *J. Biol. Chem.*, *270*, 16483-16486.

[27] Ghosh, S., May, M. J. & Kopp, E. B. (1998) NF-κB and rel proteins: evolutionarily conserved mediators of immune responses. *Annu. Rev. Immunol.*, *16*, 225-260.

[28] Baeuerle, P. A. (1998) Pro-inflammatory signaling: Last pieces in the NF-κB puzzle? *Curr. Biol.*, *8*, R19-R22.

[29] Ajuebor, M. N., Virag, L., Flower, R. J., Perretti, M. & Szabo, C. (1998) Role of inducible nitric oxide synthase in the regulation of neutrophil migration in zymosan-induced inflammation. *Immunology*, *95*, 625-630.

[30] Connelly, L., Palacios-Callender, M., Ameixa, C., Moncada, S. & Hobbs, A. J. (2001) Biphasic regulation of NF-kappaB activity underlies the pro- and anti-inflammatory actions of nitric oxide. *J. Immunol.*, *166*, 3873-3881.

[31] Peng, H.-B., Libby, P. & Liao, J. K. (1995) Induction and stabilization of IκBα by nitric oxide mediates inhibition of NF-κB. *J. Biol. Chem.*, *270*, 14214-14219.

[32] Katsuyama, K., Shichiri, M., Marumo, F. & Hirata, Y. (1998) NO inhibits cytokine-induced iNOS expression and NF-kappaB activation by interfering with phosphorylation and degradation of IkappaB-alpha. *Arterioscler. Thromb. Vasc. Biol.*, *18*, 1796-11802.

[33] Mathews, J. R., Botting, C. H., Panico, M., Morris, H. R. & Hay, R. T. (1996) Inhibition of NF-kappaB DNA binding by nitric oxide. *Nucl. Acid Res.*, *24*, 2236-2242.

[34] Pineda-Molina, E., Klatt, P., Vázquez, J., Marina, A., Lacoba, M. G. d., Pérez-Sala, D. & Lamas, S. (2001) Glutathionylation of the p50 subunit of NF-κB: a mechanism for redox-induced inhibition of DNA-binding. *Biochemistry*, *40*, 14134-14142.

[35] Shirakawa, F. & Mizel, S. (1989) In vitro activation and nuclear translocation of NF-kappaB catalyzed by cyclic AMP-dependent protein kinase and protein kinase C. *Mol. Cell. Biol.*, *9*, 2424-2430.

[36] Zhong, H., Yang, H. S., Erdjument-Bromage, H., Tempst, P. & Ghosh, S. (1997) The transcriptional activity of NF-kappaB is regulated by the IkappaB- associated PKAc subunit through a cyclic AMP-independent mechanism. *Cell*, *89*, 413-424.

[37] Poligone, B. & Baldwin, A. S. (2001) Positive and negative regulation of NF-kappaB by COX-2: roles of different prostaglandins. *J. Biol. Chem.*, *276*, 38658-38664.

[38] Nusing, R. M., Klein, T., Pfeilschifter, J. & Ullrich, V. (1996) Effect of cyclic AMP and prostaglandin $E_2$ on the induction of nitric oxide- and prostanoid-forming pathways in cultured rat mesangial cells. *Biochem. J.*, *313*, 617-623.

[39] D'Acquisto, F., Sautebin, L., Iuvone, T., Di Rosa, M. & Carnuccio, R. (1998) Prostaglandins prevent inducible nitric oxide synthase protein expression by inhibiting nuclear factor-kappaB activation in J774 macrophages. *FEBS Lett.*, *440*, 76-80.

[40] Shibata, T., Kondo, M., Osawa, T., Shibata, N., Kobayashi, M. & Uchida, K. (2002) 15-Deoxy-Delta 12,14-prostaglandin J2. A prostaglandin D2 metabolite generated during inflammatory processes. *J. Biol. Chem.*, *277*, 10459-10466.

[41] Kondo, M., Shibata, T., Kumagai, T., Osawa, T., Shibata, N., Kobayashi, M., Sasaki, S., Iwata, M., Noguchi, N. & Uchida, K. (2002) 15-Deoxy-Delta 12,14-prostaglandin J2: The endogenous electrophile that induces neuronal apoptosis. *Proc. Natl. Acad. Sci. U S A*, *99*, 7367-7372.

[42] Thomas, S. C., Ryan, M. A., Shanley, T. P. & Wong, H. R. (1998) Induction of the stress response with prostaglandin $A_1$ increases IκBα gene expression. *FASEB J.*, *12*, 1371-1378.

[43] Santoro, M. G. (2000) Heat shock factors and the control of the stress response. *Biochem. Pharmacol.*, *59*, 55-63.

[44] Forman, B. M., Tontonoz, P., Chen, J., Brun, R. P., Spiegelman, B. M. & Evans, R. M. (1995) 15-Deoxy-delta 12, 14-prostaglandin J2 is a ligand for the adipocyte determination factor PPAR gamma. *Cell*, *83*, 803-812.

[45] Kliewer, S. A., Lenhard, J. M., Willson, T. M., Patel, I., Morris, D. C. & Lehmann, J. M. (1995) A prostaglandin J2 metabolite binds peroxisome proliferator-activated receptor gamma and promotes adipocyte differentiation. *Cell*, *83*, 813-819.

[46] Schoonjans, K., Staels, B. & Auwerx, J. (1996) The peroxisome proliferator activated receptors (PPARS) and their effects on lipid metabolism and adipocyte differentiation. *Biochim. Biophys. Acta*, *1302*, 93-109.

[47] Blanquart, C., Barbier, O., Fruchart, J. C., Staels, B. & Glineur, C. (2003) Peroxisome proliferator-activated receptors: regulation of transcriptional activities and roles in inflammation. *J. Steroid Biochem. Mol. Biol.*, *85*, 267-273.

[48] Ricote, M., Li, A. C., Willson, T. M., Kelly, C. J. & Glass, C. K. (1998) The peroxisome proliferator-activated receptor-γ is a negative regulator of macrophage activation. *Nature*, *391*, 79-82.

[49] Wayman, N. S., Hattori, Y., McDonald, M. C., Mota-Filipe, H., Cuzzocrea, S., Pisano, B., Chatterjee, P. K. & Thiemermann, C. (2002) Ligands of the peroxisome proliferator-activated receptors (PPAR-gamma and PPAR-alpha) reduce myocardial infarct size. *FASEB J.*, *16*, 1027-1040.

[50] Li, M., Pascual, G. & Glass, C. K. (2000) Peroxisome proliferator-activated receptor gamma-dependent repression of the inducible nitric oxide synthase gene. *Mol. Cell. Biol.*, *20*, 4699-4707.

[51] Delerive, P., Gervois, P., Fruchart, J. C. & Staels, B. (2000) Induction of IkappaBalpha expression as a mechanism contributing to the anti-inflammatory activities of peroxisome proliferator-activated receptor-alpha activators. *J. Biol. Chem.*, *275*, 36703-36707.

[52] Delerive, P., Bosscher, K. D., Besnard, S., Berghe, W. V., Peter, J. M., Gonzalez, F. J., Fruchart, J.-C., Tedgui, A., Haegeman, G. & Staels, B. (1999) Peroxisome proliferator-activated receptor α negatively regulates the vascular inflammatory gene

response by negative cross-talk with transcription factors NF-κB and AP-1. *J. Biol. Chem.*, *274*, 32048-32054.

[53] Mendez, M. & LaPointe, M. C. (2003) PPARgamma inhibition of cyclooxygenase-2, PGE2 synthase, and inducible nitric oxide synthase in cardiac myocytes. *Hypertension*, *42*, 844-850.

[54] Fahmi, H., Di Battista, J. A., Pelletier, J. P., Mineau, F., Ranger, P. & Martel-Pelletier, J. (2001) Peroxisome proliferator-activated receptor gamma activators inhibit interleukin-1beta-induced nitric oxide and matrix metalloproteinase 13 production in human chondrocytes. *Arthritis Rheum.*, *44*, 595-607.

[55] Hortelano, S., Castrillo, A., Alvarez, A. M. & Boscá, L. (2000) Contribution of cyclopentenone prostaglandins to the resolution of inflammation through the potentiation of apoptosis in activated macrophages. *J. Immunol.*, *165*, 6525-6531.

[56] Alvarez-Maqueda, M., El Bekay, R., Alba, G., Monteseirin, J., Chacon, P., Vega, A., Martin-Nieto, J., Bedoya, F., Pintado, E. & Sobrino, F. (2004) 15-deoxy-delta 12,14-prostaglandin J2 induces heme oxygenase-1 gene expression in a reactive oxygen species-dependent manner in human lymphocytes. *J. Biol. Chem.*, *279*, 21929-21937.

[57] Noyori, R. & Suzuki, M. (1993) Organic synthesis of prostaglandins: advancing biology. *Science*, *259*, 44-45.

[58] Atsmon, J., Sweetman, B. J., Baertschi, S. W., Harris, T. M. & L. Jackson Roberts, I. (1990) Formation of thiol conjugates of 9-deoxy-$\Delta^9$,$\Delta^{12}$(*E*)-prostaglandin $D_2$ and $\Delta^{12}$(*E*)-prostaglandin $D_2$. *Biochemistry*, *29*, 3760-3765.

[59] Pérez-Sala, D., Cernuda-Morollón, E. & Cañada, F. J. (2003) Molecular basis for the inhibition of AP-1 DNA binding by 15-deoxy-$\Delta^{12,14}$-prostaglandin $J_2$. *J. Biol. Chem.*, *278*, 51251-51260.

[60] Stamatakis, K., Sánchez-Gómez, F. J. & Pérez-Sala, D. (2004) Protein modification by cyclopentenone prostaglandin addition: biological actions and therapeutic implications. *Gene Ther. Mol. Biol.*, *8*, 241-258.

[61] Straus, D. S., Pascual, G., Li, M., Welch, J. S., Ricote, M., Hsiang, C. H., Sengchanthalangsy, L. L., Ghosh, G. & Glass, C. K. (2000) 15-deoxy-delta 12,14-prostaglandin J2 inhibits multiple steps in the NF-kappaB signaling pathway. *Proc. Natl. Acad. Sci. U S A*, *97*, 4844-4849.

[62] Cernuda-Morollón, E., Pineda-Molina, E., Cañada, F. J. & Pérez-Sala, D. (2001) 15-Deoxy-$\Delta^{12,14}$-prostaglandin $J_2$ inhibition of NF-κB DNA binding through covalent modification of the p50 subunit. *J. Biol. Chem.*, *276*, 35530-35536.

[63] Parker, J. (1995) Prostaglandin $A_2$ protein interactions and inhibition of cellular proliferation. *Prostaglandins*, *50*, 359-375.

[64] Cernuda-Morollón, E., Rodríguez-Pascual, F., Klatt, P., Lamas, S. & Pérez-Sala, D. (2002) PPAR agonists amplify iNOS expression while inhibiting NF-κB: implications for mesangial cell activation by cytokines. *J. Am. Soc. Nephrol.*, *13*, 2223-2231.

[65] Appleby, S. B., Ristimaki, A., Neilson, K., Narko, K. & Hla, T. (1994) Structure of the human cyclo-oxygenase-2 gene. *Biochem. J.*, *302*, 723-727.

[66] Venema, R. C., Nishida, K., Alexander, R. W., Harrison, D. G. & Murphy, T. J. (1994) Organization of the bovine gene encoding the endothelial nitric oxide synthase. *Biochim. Biophys. Acta*, *1218*, 413-420.

[67] Pérez-Sala, D., Cernuda-Morollón, E., Pineda-Molina, E. & Cañada, F. J. (2002) Contribution of covalent protein modification to the anti-inflammatory effects of cyclopentenone prostaglandins. *Ann. N. Y. Acad. Sci.*, *973*, 533-536.

[68] Sawano, H., Haneda, M., Sugimoto, T., Inoki, K., Koya, D. & Kikkawa, R. (2002) 15-Deoxy-Delta12,14-prostaglandin J2 inhibits IL-1beta-induced cyclooxygenase-2 expression in mesangial cells. *Kidney Int.*, *61*, 1957-1967.

[69] Chatterjee, P. K., Patel, N. S., Cuzzocrea, S., Brown, P. A., Stewart, K. N., Mota-Filipe, H., Britti, D., Eberhardt, W., Pfeilschifter, J. & Thiemermann, C. (2004) The cyclopentenone prostaglandin 15-deoxy-Delta(12,14)-prostaglandin J(2) ameliorates ischemic acute renal failure. *Cardiovasc. Res.*, *61*, 630-643.

[70] Ianaro, A., Maffia, P., Cuzzocrea, S., Mazzon, E., Santoro, M. G., Di Rosa, M. & Ialenti, A. (2003) 2-Cyclopenten-1-one and prostaglandin J2 reduce restenosis after balloon angioplasty in rats: role of NF-kappaB. *FEBS Lett.*, *553*, 21-27.

[71] Ianaro, A., Ialenti, A., Maffia, P., Pisano, B. & Di Rosa, M. (2001) Role of cyclopentenone prostaglandins in rat carrageenin pleurisy. *FEBS Lett.*, *508*, 61-66.

[72] Castrillo, A., Diaz-Guerra, M. J., Hortelano, S., Martin-Sanz, P. & Bosca, L. (2000) Inhibition of IkappaB kinase and IkappaB phosphorylation by 15-deoxy- Delta(12,14)-prostaglandin J(2) in activated murine macrophages. *Mol. Cell. Biol.*, *20*, 1692-1698.

[73] Rossi, A., Kapahi, P., Natoli, G., Takahashi, T., Chen, Y., Karin, M. & Santoro, M. G. (2000) Anti-inflammatory cyclopentenone prostaglandins are direct inhibitors of IκB kinase. *Nature*, *403*, 103-108.

[74] Castrillo, A., Traves, P. G., Martin-Sanz, P., Parkinson, S., Parker, P. J. & Bosca, L. (2003) Potentiation of protein kinase C zeta activity by 15-deoxy-delta(12,14)-prostaglandin J(2) induces an imbalance between mitogen-activated protein kinases and NF-kappaB that promotes apoptosis in macrophages. *Mol. Cell. Biol.*, *23*, 1196-1208.

[75] Moos, P. J., Edes, K., Cassidy, P., Massuda, E. & Fitzpatrick, F. A. (2003) Electrophilic prostaglandins and lipid aldehydes repress redox-sensitive transcription factors p53 and hypoxia-inducible factor by impairing the selenoprotein thioredoxin reductase. *J. Biol. Chem.*, *278*, 745-750.

[76] Mullally, J. E., Moos, P. J., Edes, K. & Fitzpatrick, F. A. (2001) Cyclopentenone prostaglandins of the J series inhibit the ubiquitin isopeptidase activity of the proteasome pathway. *J. Biol. Chem.*, *276*, 30366-30373.

[77] Shibata, T., Yamada, T., Kondo, M., Tanahashi, N., Tanaka, K., Nakamura, H., Masutani, H., Yodoi, J. & Uchida, K. (2003) An endogenous electrophile that modulates the regulatory mechanism of protein turnover: inhibitory effects of 15-deoxy-Delta 12,14-prostaglandin J2 on proteasome. *Biochemistry*, *42*, 13960-13968.

[78] Levonen, A. L., Landar, A., Ramachandran, A., Ceaser, E. K., Dickinson, D. A., Zanoni, G., Morrow, J. D. & Darley-Usmar, V. M. (2004) Cellular mechanisms of redox cell signaling: the role of cysteine modification in controlling antioxidant defenses in response to electrophilic lipid oxidation products. *Biochem. J.*, *378*, 373-382.

[79] Itoh, K., Mochizuki, M., Ishii, Y., Ishii, T., Shibata, T., Kawamoto, Y., Kelly, V., Sekizawa, K., Uchida, K. & Yamamoto, M. (2004) Transcription Factor Nrf2

Regulates Inflammation by Mediating the Effect of 15-Deoxy-Delta(12,14)-Prostaglandin J(2). *Mol. Cell. Biol.*, *24*, 36-45.

[80] Shibata, T., Yamada, T., Ishii, T., Kumazawa, S., Nakamura, H., Masutani, H., Yodoi, J. & Uchida, K. (2003) Thioredoxin as a molecular target of cyclopentenone prostaglandins. *J. Biol. Chem.*, *278*, 26046-26054.

[81] Lee, J. H., Koo, T. H., Hwang, B. Y. & Lee, J. J. (2002) Kaurane diterpene, kamebakaurin, inhibits NF-kappaB by directly targeting the DNA-binding activity of p50 and blocks the expression of antiapoptotic NF-kappaB target genes. *J. Biol. Chem.*, *277*, 18411-18420.

[82] Perry, N. B., Albertson, G. D., Blunt, J. W., Cole, A. L., Munro, M. H. & Walker, J. R. (1991) 4-Hydroxy-2-cyclopentenone: an anti-Pseudomonas and cytotoxic component from Passiflora tetrandra. *Planta Med.*, *57*, 129-131.

[83] Vcrbitski, S. M., Mullally, J. E., Fitzpatrick, F. A. & Ireland, C. M. (2004) Punaglandins, chlorinated prostaglandins, function as potent Michael receptors to inhibit ubiquitin isopeptidase activity. *J. Med. Chem.*, *47*, 2062-2070.

[84] Fenaille, F., Tabet, J.-C. & Guy, P. A. (2004) Identification of 4-hydroxy-2-nonenal-modified peptides within unfractionated digests using matrix-assisted laser desorption/ionization time-of-flight mass spectrometry. *Anal. Chem.*, *76*, 867-873.

[85] Burcham, P. C., Fontaine, F. R., Kaminskas, L. M., Petersen, D. R. & Pyke, S. M. (2004) Protein adduct-trapping by hydrazinophthalazine drugs: mechanisms of cytoprotection against acrolein-mediated toxicity. *Mol. Pharmacol*, *65*, 655-664.

[86] Mahon, T. M. & O'Neill, L. A. (1995) Studies into the effect of the tyrosine kinase inhibitor herbimycin A on NF-kappaB activation in T lymphocytes. Evidence for covalent modification of the p50 subunit. *J. Biol. Chem.*, *270*, 28557-28564.

[87] Ogino, S., Tsuruma, K., Uehara, T. & Nomura, Y. (2004) Herbimycin A abrogates Nuclear Factor-kappaB activation by interacting preferentially with the IkappaB Kinase beta subunit. *Mol. Pharmacol.*, *65*, 1344-1351.

In: Trends in DNA Research
Editor: Corey R. Woods, pp. 33-51
ISBN 1-59454-635-5


*Chapter II*

# Overview of Longevity-Associated Mitochondrial DNA 5178 C/A Polymorphism and a Discussion of its Modulation of the Effects of Habitual Smoking on Serum Total and LDL Cholesterol Levels in Middle-aged Japanese Men

*Akatsuki Kokaze,*[a*] *Mamoru Ishikawa,*[a,b] *Naomi Matsunaga,*[a] *Masao Yoshida,*[a] *Ryuji Makita,*[c] *Masao Satoh,*[d] *Koji Teruya,*[e] *Kanako Sekiguchi,*[a] *Yumi Masuda,*[a] *Matsuko Harada,*[a] *Yoshiko Uchida*[a] *and Yutaka Takashima*[a]

[a]Department of Public Health, Kyorin University School of Medicine, Mitaka-shi, Tokyo, Japan
[b]Mito Red Cross Hospital, Mito-shi, Ibaraki, Japan
[c]Department of General Medicine, Kyorin University School of Medicine, Mitaka-shi, Tokyo, Japan
[d]Department of Allergy and Immunology, Saitama Medical School, Moroyama, Iruma-gun, Saitama, Japan
[e]Department of Public Health, Kyorin University School of Health Sciences, Hachioji-shi, Tokyo, Japan

* Correspondence to: Akatsuki Kokaze, Department of Public Health, Kyorin University School of Medicine, 6-20-2, Shinkawa, Mitaka-shi, Tokyo, 181-8611, Japan. Tel: +81(422)47-5511 Ext 3495; Fax: +81(422)76-0366; E-mail: dawn@kyorin-u.ac.jp

# Abstract

Mitochondrial DNA 5178 cytosine/adenine (Mt5178 C/A) polymorphism, which is also known as NADH dehydrogenase subunit-2 237 leucine/methionine (ND2-237 Leu/Met) polymorphism, is reportedly associated with longevity in the Japanese population, and the Mt5178A genotype may possess antiatherogenic properties. Individuals with the Mt5178A genotype may be more resistant to atherogenic diseases, such as myocardial infarction and cerebrovascular diseases, than those with Mt5178C. Moreover, Mt5178 C/A polymorphism has been reported to be associated with blood pressure, intraocular pressure, serum protein fraction levels, and serum lipid levels in healthy Japanese individuals.

In order to investigate the interaction between Mt5178 C/A polymorphism and habitual smoking on serum lipid levels, we performed a cross-sectional study in which healthy Japanese male subjects were classified into two groups based on body mass index (BMI): BMI $\leq$ 22 (n = 107); and BMI > 22 (n = 214). Interaction between Mt5178 C/A polymorphism and cigarette smoking was observed with regard to serum triglyceride (TG) levels in BMI $\leq$ 22 subjects (P = 0.026), and with regard to serum total cholesterol (TC) levels and serum low-density lipoprotein cholesterol (LDLC) levels in BMI > 22 subjects (P = 0.019 and P = 0.043, respectively). In BMI > 22 subjects, TC levels were significantly lower in non-smokers with Mt5178A than in smokers with Mt5178A (P = 0.015), while LDLC levels were lower in non-smokers with Mt5178A than in those with Mt5178C or in smokers with Mt5178A (P = 0.047 and P = 0.055, respectively). Although interaction between Mt5178 C/A polymorphism and habitual smoking was not observed with regard to TG levels in BMI > 22 subjects (P = 0.308), TG levels were significantly higher in smokers with Mt5178A than in non-smokers with Mt5178A as well as in non-smokers with Mt5178C (P = 0.011 and P = 0.014, respectively). Taken together, these results suggest that habitual smoking may deprive Japanese Mt5178A (ND2-237Met) genotypic men having BMI of > 22 of their antiatherogenic advantage.


**Keywords:** Mitochondrial DNA 5178 cytosine/adenine polymorphism; NADH dehydrogenase; longevity; habitual smoking; serum total cholesterol levels; serum low-density lipoprotein cholesterol levels

# Introduction

## Mitochondrial DNA 5178 C/A polymorphism

Numerous studies on mitochondria have revealed their cell biology, molecular genetics, energy metabolism and mitochondria-related diseases, as well as their application to anthropology and forensic science [1]. Over the past few decades, a considerable number of studies have been conducted regarding the association of mitochondria with aging and longevity [2, 3]. As a result of these studies, several mitochondrial DNA polymorphisms have been reported to be associated with longevity [4-9]. Mitochondrial DNA 5178 cytosine/adenine (Mt5178 C/A) polymorphism, which results in NADH dehydrogenase subunit-2 237 leucine/methionine (ND2-237 Leu/Met) polymorphism, is known to be one of these longevity-associated mitochondrial DNA polymorphisms.

Tanaka and his colleagues, who performed a comprehensive sequence analysis of the entire mitochondrial genome in centenarians, both obese and non-obese young adults, patients with Parkinson's or Alzheimer's disease and diabetic patients [10], first suggested that Mt5178 C/A (ND2-237 Leu/Met) polymorphism was associated with longevity in Japanese people [4, 5]. They reported that the frequency of Mt5178A (ND2-237Met), representing mitochondrial DNA haplogroup D, was significantly higher in centenarians than in the general population. The Mt5178A genotype is generally rare worldwide [11, 12], and is believed to be as high as 40-45% in the Japanese populace [4, 13-21]. There have been a number of reports regarding the association between Mt5178 C/A polymorphism and adult-onset diseases [14, 16, 17, 22], while individuals with Mt5178A appear to be resistant to atherosclerotic diseases. These observations on Mt5178 C/A polymorphism and its association with life expectancy may help to explain why life expectancy in Japan is one of the longest in the world.

Wang et al. investigated the association of longevity-associated Mt5178 C/A polymorphism with the occurrence and clinical features of type 2 diabetes [14]. The frequency of the Mt5178A genotype was lower in patients with type 2 diabetes than in controls. Moreover, patients with a maternal history of diabetes tended to carry the Mt5178C genotype more frequently when compared to those without a maternal history of diabetes and controls. Mean age at onset of diabetes was significantly lower in patients with Mt5178C than in those with Mt5178A, while insulin treatment tended to be more frequent in Mt5178C genotypic patients. Matsunaga et al. investigated the significance of Mt5178 C/A polymorphism in type 2 diabetic subjects [15]. Using ultrasonography, they assessed the association between the Mt5178 C/A genotypes and early atherosclerotic changes of the bilateral carotid arteries. Mean intima-media thickness in the bilateral arteries was significantly greater in the Mt5178C genotypic group than in the Mt5178A genotypic group, and the Mt5178 C/A polymorphism was significantly correlated with mean intima-media thickness and the presence of plaques. They concluded that the Mt5178A genotype exerted an antiatherogenic effect in type 2 diabetic patients.

Mukae et al. investigated the relationship between longevity-associated Mt5178 C/A polymorphism and acute myocardial infarction [16]. They reported that the frequency of the Mt5178C genotype was significantly higher in patients with acute myocardial infarction than in control subjects, and that this difference was more evident in younger patients. They concluded that Mt5178C may promote the occurrence of acute myocardial infarction, and that the Mt5178C genotype was possibly a novel coronary risk factor in Japanese. Takagi et al. also reported that the frequency of Mt5178A genotype was significantly higher in controls than in subjects with myocardial infarction after adjustment for age, gender, body mass index (BMI), smoking status, hypertension, diabetes, hypercholesterolemia and hyperuricemia [22].

Aging is one of the risk factors associated with cerebrovascular diseases. It is therefore epidemiologically and etiologically interesting to investigate the correlation between longevity-associated Mt5178 C/A polymorphism and cerebrovascular diseases. Ohkubo et al. investigated Mt5178 C/A polymorphism and angiotensin I-converting enzyme gene polymorphism in patients with cerebrovascular disorder and age-matched normal controls

[17]. Although there was no significant difference in the frequencies of angiotensin I-converting enzyme genotype between the cerebrovascular disorder patients and controls, the frequency of Mt5178C was significantly higher in cerebrovascular disorder patients than in normal controls. This report thus suggests that the Mt5178C genotype is one of the genetic risk factors for cerebrovascular diseases in Japanese individuals.

Interestingly, Aoyama et al. conducted a cross-sectional analysis of Trp64Arg polymorphism in the $\beta_3$-adrenergic receptor gene (BAR-3) and Mt5178 C/A (ND2-237 Leu/Met) polymorphism in order to determine whether these polymorphisms are correlated with lifestyle-related phenotypes and nutrient intake [18]. They showed that body fat content was significantly higher in subjects with the BAR-3 variant genotype (Arg/Arg or Arg/Trp) and Mt5178A when compared to those with the BAR-3 normal genotype (Trp/Trp) and Mt5178A, and those with BAR-3 normal genotype and Mt5178C. Carbohydrate energy ratio was significantly higher and animal protein intake was significantly lower in subjects with the BAR-3 normal genotype and Mt5178A than in those with the other three genotypic combinations. Combinations of BAR-3 polymorphism and Mt5178 C/A polymorphism therefore appear to have an influence on eating behavior. Moreover, Kato et al. reported that Mt5178 C/A (ND2-237 Leu/Met) polymorphism is associated with bipolar disorder [19]; individuals with Mt5178C are reportedly more susceptible to bipolar disorder than those with Mt5178A.

The leading cause of death in Japan is cancer, and the question of whether Mt5178 C/A polymorphism is associated with cancer risk is of particular interest. Iwao et al. reported no association between Mt5178 C/A polymorphism and six types of cancer (esophageal cancer, stomach cancer, colorectal cancer, lung cancer, breast cancer and malignant lymphoma) in the Japanese population [20]. However, they also reported that the interaction between Mt5178 C/A genotype, age and cancer, was significant in lung cancer subjects aged $\geq 60$ years. Therefore, further investigation is required into the association between longevity-associated Mt5178 C/A polymorphism and susceptibility to cancer.

As discussed above, Mt5178 C/A polymorphism is widely associated with the pathologies of common diseases. In fact, these observations prompted us to investigate whether Mt5178 C/A polymorphism may be associated with biophysical and biochemical factors in healthy middle-aged subjects. Therefore, with a focus on preventive medicine or successful aging, we conducted cross-sectional studies on the association of Mt5178 C/A polymorphism with blood pressure [23], intraocular pressure [24], serum protein fraction levels [25, 26] and serum lipid levels [21, 27] in primarily healthy subjects.

## Mt5178 C/A Polymorphism and Blood Pressure

We previously reported that Mt5178 C/A (ND2-237 Leu/Met) polymorphism may influence both diastolic blood pressure (DBP) in Japanese women and the blood-pressure-increasing effect of habitual drinking in Japanese men [23]. In healthy Japanese women, DBP was significantly higher in those with the Mt5178A (ND2-237Met) genotype ($n = 41$; mean DBP ± SE. = 70.2 ± 1.2 mmHg) than in those with the Mt5178C (ND2-237Leu) genotype ($n = 50$; 67.0 ± 1.0 mmHg) ($P = 0.040$). After adjusting for age, BMI, habitual drinking, and habitual smoking, this difference also remained present ($P = 0.022$). In healthy Japanese men, Mt5178 C/A polymorphism was not found to be associated with blood pressure. However, the results of thorough data analysis suggested that Mt5178 C/A polymorphism may be indirectly associated with systolic blood pressure (SBP) and DBP via drinking frequency. After adjusting for age, BMI, and habitual smoking, SBP was revealed to be significantly higher in daily drinkers with the Mt5178C genotype ($n = 81$; least-square mean SBP ± SE. = 128.3 ± 1.5 mmHg) than in occasional drinkers ($n = 43$; 120.3 ± 2.1 mmHg) and non-drinkers ($n = 69$; 118.7 ± 1.6 mmHg) with the Mt5178C genotype ($P = 0.002$ and $P < 0.001$, respectively). Moreover, DBP was observed to be significantly higher in daily drinkers with Mt5178C ($n = 81$; least-square mean DBP ± SE. = 75.1 ± 1.0 mmHg) than in occasional drinkers ($n = 43$; 69.9 ± 1.4 mmHg) and non-drinkers ($n = 69$; 70.1 ± 1.1 mmHg) with Mt5178C ($P = 0.002$ and $P < 0.001$, respectively). However, in men with Mt5178A, no significant differences in blood pressure among daily drinkers, occasional drinkers and non-drinkers were observed. Therefore, this may be a novel gene-environmental (lifestyle) interaction on blood pressure.

The mechanism of the interaction described above has proven to be difficult fully elucidate. NADH dehydrogenase, complex I in the respiratory chain, is involved in ethanol-related production of reactive oxygen species (ROS) [28]. Superoxide radicals in and around vascular endothelial cells play a critical role in the pathogenesis of hypertension [29]. Mitochondria reportedly have a biophysical influence on blood pressure [30-32]. Therefore, it has been postulated that biochemical differences in the function of the ND2-237Leu and ND2-237Met isoforms of NADH dehydrogenase are involved in the interaction of ND2-237 Leu/Met polymorphism with habitual drinking on blood pressure. However, such a hypothesis requires further investigation.

## Mt5178 C/A Polymorphism and Intraocular Pressure

We also reported that Mt5178 C/A polymorphism may be associated with intraocular pressure (IOP) in middle-aged Japanese men [24]. Because age-related changes in IOP are associated with ethnicity [33-36] and/or atherosclerosis [37, 38], it is interesting from a biophysical standpoint to analyze the association between longevity-associated Mt5178 C/A polymorphism and IOP. In one such study, 386 Japanese men, who had not previously undergone any medical treatment for hypertension, ocular hypertension or glaucoma, were selected as subjects. Mean IOP was significantly higher in men with Mt5178C ($n = 234$; mean IOP ± SE. = 13.3 ± 0.2 mmHg) than in those with Mt5178A ($n = 152$; 12.7 ± 0.2 mmHg) ($P = 0.037$). This difference in mean IOP between Mt5178 C/A genotypes remained evident after adjusting for age, BMI, blood pressure, habitual smoking and habitual drinking. In addition, interactions between Mt5178 C/A polymorphism and habitual smoking or habitual drinking with regard to IOP were observed. Multiple regression analysis revealed that habitual smoking was significantly associated with IOP in men with Mt5178A ($P$ for trend = 0.020), while daily alcohol consumption was significantly associated with IOP in those with Mt5178C ($P$ for trend = 0.021).

Mitochondria are reportedly involved in the regulation of IOP [39-41]. Such reports demonstrate the need for further investigation to elucidate the mechanisms of association between Mt5178 C/A polymorphism and IOP.

## Mt5178 C/A and Serum Protein Fraction Levels

Age-related changes in serum protein, albumin, and globulin levels have been reported [42-44], and thus the association between longevity-associated Mt5178 C/A polymorphism and serum protein fraction levels was examined. Alpha-1, alpha-2, and beta globulin proportions were significantly higher in Mt5178A genotypic women ($n$ = 44) than in Mt5178C genotypic women ($n$ = 55) ($P$ = 0.002, $P$ = 0.006, and $P$ = 0.008, respectively) [25]. Moreover, alpha-1, alpha-2, and beta globulin levels were significantly higher in Mt5178A genotypic women than in Mt5178C genotypic women ($P$ = 0.001, $P$ = 0.002, and $P$ = 0.018, respectively). These differences in serum globulin levels between the Mt5178 C/A genotypes were more evident in premenopausal women than in postmenopausal women. However, no such differences in serum protein fraction levels were observed in men. Tobacco smoke [45-47], even environmental tobacco smoke [48], influences serum protein fraction levels. We therefore investigated whether there was a difference in the effects of habitual smoking on serum protein fraction levels in Mt5178 C/A genotypic men [26]. In Mt5178C genotypic men, alpha-1 and alpha-2 globulin levels were significantly higher in smokers than in non-smokers ($P < 0.001$ and $P$ = 0.002, respectively), and the influence of smoking on these globulin levels depended on daily cigarette consumption. However, in men with the Mt5178A genotype, no significant differences were observed in alpha-1 or alpha-2 globulin levels between smokers and non-smokers.

Levels of alpha-1 antitrypsin, alpha-1 acid glycoproteins, and alpha-2 macroglobulin are reported to be strongly associated with degree of atherosclerosis [49]. These findings suggest that atherosclerosis may be more severe in smokers with Mt5178C than in non-smokers with Mt5178C, and Mt5178A genotypic men may be resistant to the aggravating effects of smoking on atherosclerosis. However, another point needs to be considered. Alpha-1 and alpha-2 globulin levels were significantly higher in premenopausal women with Mt5178A than in those with Mt5178C. Both alpha-1 antitrypsin and alpha-1 acid glycoproteins are electrophoretically related to alpha-1 globulin. Hashimoto et al. reported that alpha-1 antitrypsin levels were lower after menopause and were increased by estrogen, while alpha-1 acid glycoproteins were higher after menopause and were decreased by estrogen [50]. Denko and Gabriel reported that alpha-1 antitrypsin exhibited a biphasic change during the aging process, while alpha-1 acid glycoproteins decreased linearly as age increased [51]. Therefore, further investigation is required in order to confirm the differences in serum protein levels between young and old individuals and those between men and women.

### Mt5178 C/A Polymorphism and Serum Lipid Levels

We previously reported that Mt5178 C/A polymorphism may be associated with serum lipid levels in the Japanese population [21]. In healthy Japanese men, after adjusting for age and BMI, high-density lipoprotein cholesterol (HDLC) levels were significantly higher in those with the Mt5178A genotype ($n$ = 148; mean HDLC ± SE. = 57.8 ± 1.2 mg/dl) than in those with the Mt5178C genotype ($n$ = 214; 54.4 ± 1.0 mg/dl) ($P$ = 0.026). However, this significant difference disappeared after adjusting for drinking frequency. We therefore investigated the interaction between Mt5178 C/A polymorphism and habitual drinking on serum lipid levels in 321 healthy middle-aged Japanese men [27]. Interaction between Mt5178 C/A polymorphism and daily drinking on serum triglyceride (TG) levels was observed ($P$ = 0.019). Moreover, interaction between Mt5178 C/A polymorphism and cigarette consumption on serum TG levels was also observed ($P$ = 0.022). Multiple regression analysis showed that in Mt5178A genotypic men, daily drinking decreased TG levels ($P$ = 0.025) and that cigarette consumption increased TG levels ($P < 0.001$), while in Mt5178C genotypic men, the effects of daily drinking and cigarette consumption on TG levels were unclear.

In healthy Japanese women, TG levels were significantly lower in those with the Mt5178A genotype ($n$ = 44; 84.5 ± 6.8 mg/dl) than in those with the Mt5178C genotype ($n$ = 55; 108.1 ± 6.1 mg/dl) ($P$ = 0.012) [21]. This difference in TG levels between the Mt5178 C/A genotypes was more evident in postmenopausal women than in premenopausal women.

### Aim of this Study

As discussed above, the antiatherogenic effects of the Mt5178A genotype have been reported in healthy subjects [21, 26] and in type 2 diabetic patients [15]. We have also reported interactions between Mt5178 C/A polymorphism and cigarette smoking on serum TG levels [27], IOP [24], serum protein fraction levels [26] and red blood cell counts [52] in Japanese men. Habitual smoking is a major risk factor for atherosclerosis [53], and is harmful to mitochondria [54]. Moreover, body mass index (BMI) levels appeared to strongly influence the relationship between cigarette smoking and serum lipid levels in a previous study [27]. Therefore, in order to investigate whether Mt5178 C/A (ND2-237 Leu/Met) polymorphism modulates the effects of cigarette smoking on serum lipid levels, we performed a cross-sectional study in which Japanese male subjects were classified into two groups based on BMI: BMI $\leq 22$; and BMI $> 22$.

## Subjects and Methods

### Subjects

A total of 602 volunteers were recruited from among individuals visiting the Mito Red Cross Hospital for regular medical check-ups. This study was conducted in accordance with the Declaration of Helsinki (Edinburgh 2000) and was approved by the Ethical Committee of Kyorin University School of Medicine. Written informed consent was obtained from all volunteers before participation. Among the volunteers, 327 men were enrolled in this study, and these men did not have any chronic diseases or take antihypertensive, antihyperlipideamic or antidiabetic medications. However, 6 individuals with unclear drinking frequency were subsequently excluded from analysis. Therefore, subjects of this study comprised 321 Japanese men (age, 53.0 ± 7.7 y; mean ± SD).

## Clinical Characteristics of Subjects

Determination of blood chemical and physical data was conducted as described previously [21]. Briefly, venous blood was drawn after a minimum fasting period of 12 h. Serum total cholesterol levels (TC) were measured by Determiner L TC II (Kyowa Medex) [55], serum HDLC levels by Determiner L HDL-C (Kyowa Medex) [56], and serum TG levels by Determiner L TG II (Kyowa Medex) [57]. Serum low-density lipoprotein cholesterol (LDLC) levels were calculated according to Friedewald's formula [58]. Body mass index was defined as the ratio of subject weight (kg) to the square of subject height (m). A survey of drinking and smoking habits was performed by means of a questionnaire. Habitual drinking was classified based on drinking frequency (daily drinkers; occasional drinkers, which included those who drink several times per week; and non- or ex-drinkers, which include those who drink a few times per month). Habitual smoking was classified into non- or ex-smokers and current smokers.

## Mt5178 C/A Genotyping

DNA was extracted from whole blood using the DNA Extractor WB kit (Wako Pure Chemical Industries, Japan). The forward and reverse oligonucleotide primers used to determine the Mt5178 C/A genotype were: 5'-CTTAGCATACTCCTCAATTACCC-3' and 5'-CTGAATTCTTCGATAATGGCCCA-3'. Polymerase chain reaction (PCR) was performed with 50 ng genomic DNA in buffer containing 1.5 mM $MgCl_2$, 1.25 mM dNTPs and 0.5 μl Amplitaq DNA polymerase (GeneAmp, Perkin Elmer, USA). After, initial denaturation at 94°C for 5 min, PCR was conducted through 40 cycles as follows: denaturation at 94°C for 30 s, annealing at 60°C for 60 s and polymerase extension at 72°C for 10 min. The PCR product was then digested with restriction enzyme *Alu*I (Nippon Gene, Japan) and electrophoresed on 1.5% agarose gels. Gels were stained with ethidium bromide, visualized under UV light, and photographed. The absence of an *Alu*I site was designated as Mt5178A (ND2-237Met), and the presence of this restriction site was designated as Mt5178C (ND2-237Leu).

## Statistical Analyses

For all analyses, subjects were divided into two groups based on BMI: "BMI ≤ 22" (n = 107); and "BMI > 22" (n = 214). The cut-off point (BMI = 22) was determined based on reports about optimal BMI [59]. Statistical analyses were performed using SAS statistical software, version 6.12 for Windows (SAS Institute, Inc., Cary, NC, 1997). In analysis of covariates (ANCOVA), the Mt5178 C/A genotypes (Mt5178C = 0, Mt5178A = 1), habitual smoking (non- or ex-smokers = 0, current smokers = 1), and habitual drinking (non-drinkers = 0, occasional-drinkers = 1, daily-drinkers = 2) were numerically coded. Age and BMI were included as covariates in the models. *P* values of < 0.05 were considered statistically significant.

# Results

The frequency of the Mt5178A (ND2-237Met) genotype was 40.2% (43/107) in the BMI ≤ 22 group and 39.7% (85/214) in the BMI > 22 group. No significant differences in Mt5178A genotypic frequency were observed between the two BMI-classified groups (*P* = 0.936). No significant differences in baseline characteristics were observed between the Mt5178 C/A genotypes in either the BMI ≤ 22 or BMI > 22 group (Table 1).

**Table 1. Characteristics and serum lipid levels by genotype in the BMI≤ 22 and BMI > 22 groups**

| | BMI ≤ 22 | | | BMI > 22 | | |
|---|---|---|---|---|---|---|
| | Mt51878C (n = 64) | mt5178A (n = 43) | *P* value | Mt5178C (n = 129) | Mt5178A (n = 85) | *P* value |
| Age (y) | 53.38 ± 1.04 | 52.95 ± 1.27 | 0.798 | 53.36 ± 0.65 | 52.22 ± 0.80 | 0.273 |
| BMI (kg/m$^2$) | 20.19 ± 0.17 | 20.58 ± 0.21 | 0.158 | 24.58 ± 0.17 | 24.84 ± 0.21 | 0.342 |
| Total cholesterol (mg/dl) | 203.23 ± 4.04 | 200.12 ± 4.93 | 0.626 | 203.16 ± 3.07 | 202.66 ± 3.78 | 0.918 |
| HDL cholesterol (mg/dl) | 60.28 ± 2.22 | 65.00 ± 2.71 | 0.181 | 51.67 ± 1.08 | 53.33 ± 1.33 | 0.333 |
| LDL cholesterol (mg/dl) | 120.59 ± 3.92 | 113.87 ± 4.78 | 0.279 | 124.13 ± 2.96 | 121.48 ± 3.65 | 0.574 |
| Triglyceride (mg/dl) | 111.83 ± 7.27 | 106.24 ± 8.87 | 0.627 | 136.85 ± 5.70 | 139.25 ± 7.02 | 0.791 |

Age, body mass index (BMI), total cholesterol levels, high-density lipoprotein (HDL) cholesterol levels, low-density lipoprotein (LDL) cholesterol levels, and triglyceride levels are given as mean ± SE. All *P* values show the significance of differences between Mt5178 C/A genotypes. LDL cholesterol levels were calculated according to the Friedewald formula [58].

ANCOVA revealed interactions in the BMI ≤ 22 group (Table 2) between Mt5178 C/A (ND2-237 Leu/Met) polymorphism and habitual smoking with regard to TG levels (*P* = 0.026). Interaction between Mt5178 C/A and habitual drinking was also observed with regard to TC and TG levels (*P* = 0.028 and *P* = 0.046, respectively). Habitual drinking was significantly associated with TC, HDLC and LDLC levels (*P* = 0.027, *P* = 0.002, and *P* < 0.001, respectively), and BMI was significantly associated with TC, LDLC, and TG levels (*P* = 0.049, *P* = 0.039, and *P* = 0.006, respectively).

ANCOVA revealed interactions in the BMI > 22 group (Table 3) between Mt5178 C/A polymorphism and habitual smoking with regard to TC and LDLC levels ($P = 0.019$ and $P = 0.043$, respectively). Moreover, habitual smoking was significantly associated with TG levels ($P = 0.005$), habitual drinking was only significantly associated with HDLC levels ($P < 0.001$), and BMI was significantly associated with HDLC and TG levels ($P < 0.001$ and $P = 0.006$, respectively).

Table 4 shows the least square mean levels of TC, HDLC, LDLC and TG levels after adjusting for age, BMI and habitual drinking according to Mt5178 C/A polymorphism and smoking activity in both the BMI ≤ 22 and BMI > 22 groups. In subjects with BMI of ≤ 22, no significant differences were observed in serum lipid levels between the Mt5178 C/A genotypes. Although the differences did not reach significance, TG levels were lower in smokers with Mt5178C than in non-smokers with Mt5178C ($P = 0.124$), and higher in smokers with Mt5178A than non-smokers with Mt5178C ($P = 0.105$). In subjects with BMI of > 22, TC levels were significantly higher in smokers with Mt5178A than in non-smokers with Mt5178A ($P = 0.015$). In addition, LDLC levels were significantly lower in non-smokers with Mt5178A than in those with Mt5178C ($P = 0.047$), while these levels were higher, but not significantly so, in smokers with Mt5178A when compared to non-smokers with Mt5178A ($P = 0.055$).

**Table 2. Analysis of covariance for serum lipid levels in subjects BMI of ≤ 22**

| | Total cholesterol F value | HDL cholesterol F value | LDL cholesterol F value | Triglyceride F value |
|---|---|---|---|---|
| Mt5178 C/A | 0.22 | 0.78 | 0.60 | 0.42 |
| Age | 0.02 | 0.15 | 0.53 | 0.87 |
| BMI | 3.96* | 3.17 | 4.40* | 7.81* |
| Habitual smoking | 0.01 | 3.06 | 1.44 | 0.02 |
| Habitual drinking | 3.76* | 6.79** | 8.75*** | 1.09 |
| Mt5178 C/A × habitual smoking | 0.05 | 0.22 | 0.13 | 5.12* |
| Mt5178 C/A × habitual drinking | 3.71* | 0.40 | 1.27 | 3.18* |

HDL: high-density lipoprotein.; LDL: low-density lipoprotein.; BMI: body mass index. *$P < 0.05$, **$P < 0.005$, ***$P < 0.001$. In ANCOVA, some independent variables were numerically coded: Mt5178 C/A genotypes (Mt5178C = 0, Mt5178A = 1), habitual drinking (non- or ex-drinkers = 0, occasional-drinkers = 1, daily drinkers = 2), and habitual smoking (non- or ex-smokers = 0, current smokers = 1). Mt5178 C/A × smoking habit represents interaction between Mt5178 C/A genotypes and habitual smoking. Mt5178 C/A × habitual drinking represents interaction between Mt5178 C/A genotypes and habitual drinking.

## Discussion

The present study demonstrates the interaction between Mt5178 C/A (ND2-237 Leu/Met) polymorphism and cigarette smoking on serum lipid levels in middle-aged Japanese men. Significant interactions between Mt5178 C/A polymorphism and habitual smoking with regard to TG levels were observed in "BMI ≤ 22" subjects (Table 2). In "BMI > 22" subjects, interactions between Mt5178 C/A polymorphism and habitual smoking with regard to TC and LDLC levels were observed (Table 3); TC and LDLC levels were significantly higher in smokers than in non-smokers among Mt5178A (ND2-237Met) genotypic men, while among Mt5178C (ND2-237Leu) genotypic men, neither of these differed significantly between non-smokers and smokers (Table 4).

**Table 3. Analysis of covariance for serum lipid levels in subjects with BMI of > 22**

| | Total cholesterol F value | HDL cholesterol F value | LDL cholesterol F value | Triglyceride F value |
|---|---|---|---|---|
| Mt5178 C/A | 0.08 | 1.19 | 0.62 | 0.09 |
| Age | 0.31 | 0.30 | 0.17 | 0.00 |
| BMI | 2.49 | 15.27*** | 3.72 | 7.76* |
| Habitual smoking | 1.00 | 2.86 | 0.30 | 8.06** |
| Habitual drinking | 1.38 | 9.29*** | 2.10 | 1.01 |
| Mt5178 C/A × habitual smoking | 5.55* | 0.01 | 4.14* | 1.05 |
| Mt5178 C/A × habitual drinking | 1.00 | 2.30 | 1.60 | 1.27 |

HDL: high-density lipoprotein.; LDL: low-density lipoprotein.; BMI: body mass index. *$P < 0.05$, **$P < 0.005$, ***$P < 0.001$. In ANCOVA, some independent variables were numerically coded: Mt5178 C/A genotypes (Mt5178C = 0, Mt5178A = 1), habitual drinking (non- or ex-drinkers = 0, occasional-drinkers = 1, daily drinkers = 2), and habitual smoking (non- or ex-smokers = 0, current smokers = 1). Mt5178 C/A × smoking habit represents interaction between Mt5178 C/A genotypes and habitual smoking. Mt5178 C/A × habitual drinking represents interaction between Mt5178 C/A genotypes and habitual drinking.

**Table 4. Serum lipid levels in non-smokers and smokers by genotype**

| | Mt5178C | | Mt5178A | |
|---|---|---|---|---|
| **BMI ≤ 22** | Non-smokers (n = 34) | Smokers (n = 30) | Non-smokers (n = 21) | Smokers (n = 22) |
| Total cholesterol (mg/dl) | 202.72 ± 5.37 | 202.32 ± 6.30 | 200.50 ± 6.77 | 202.84 ± 6.66 |
| HDL cholesterol (mg/dl) | 64.10 ± 2.94 | 59.53 ± 3.45 | 69.19 ± 3.70 | 61.47 ± 3.65 |
| LDL cholesterol (mg/dl) | 114.99 ± 5.02 | 123.62 ± 5.89 | 112.98 ± 6.33 | 117.49 ± 6.24 |
| Triglyceride (mg/dl) | 118.13 ± 9.62 | 95.84 ± 11.28 | 91.68 ± 12.13 | 119.50 ± 11.94 |
| **BMI > 22** | Non-smokers (n = 71) | Smokers (n = 58) | Non-smokers (n = 55) | Smokers (n = 30) |
| Total cholesterol (mg/dl) | 205.17 ± 4.19 | 201.09 ± 4.63 | 192.92 ± 4.92 | 212.26 ± 6.53† |
| HDL cholesterol (mg/dl) | 52.66 ± 1.39 | 49.81 ± 1.53 | 53.10 ± 1.63 | 50.51 ± 2.16 |
| LDL cholesterol (mg/dl) | 126.96 ± 4.03 | 121.97 ± 4.46 | 114.47 ± 4.72* | 129.03 ± 6.28 |
| Triglyceride (mg/dl) | 128.27 ± 7.65 | 146.55 ± 8.45 | 126.72 ± 8.97 | 163.55 ± 11.92*,† |

Total cholesterol, high-density lipoprotein (HDL) cholesterol, low-density lipoprotein (LDL) cholesterol, and triglyceride levels are given as least-square mean ± SE adjusted for age, body mass index and habitual drinking. *$P < 0.05$ vs non-smokers with mt5178C; †$P < 0.05$ vs non-smokers with mt5178A.

The Mt5178A genotype has been reported to exert antiatherogenic effects [15, 21, 26]. Individuals with Mt5178C were found to be more susceptible to not only type 2 [14], but also type 1 diabetes [60]. Individuals with Mt5178A are reportedly more resistant to atherogenic diseases, such as myocardial infarction [16, 22] and cerebrovascular disorders [17]. However, the present results and the fact that high levels of LDLC are epidemiologically known to be a risk factor for atherosclerosis [53] suggest that habitual smoking may counteract the antiatherogenic effects of Mt5178A (ND2-237Met).

Epidemiological approaches have demonstrated the relationship of BMI and smoking with serum lipid levels in Japanese men [61, 62]. In middle-aged Japanese men, BMI is positively associated with LDLC and TG levels, and negatively associated with HDLC levels, while habitual smoking is positively associated with TG levels, and negatively associated with HDLC levels [61]. Our findings indicate the involvement of Mt5178 C/A (ND2-237 Leu/Met) polymorphism in these relationships. In addition, interactions between genetic polymorphism and cigarette smoking on serum lipid levels have recently been reported [63-65]. Interaction between lipoprotein lipase-*Hind*III polymorphism, apolipoprotein E polymorphism and habitual smoking (gene-gene-environmental interaction) was reported with regard to HDLC levels [64]. It therefore appears that for personalized prevention of lifestyle-related atherogenic diseases, consideration of individual genetic backgrounds may possibly be more effective than conventional approaches. Activity of cholesteryl ester transfer protein (CETP), which is associated with *Taq*IB polymorphism in the CETP gene, is positively associated with LDLC levels in the Japanese population [66], while polymorphism in the intestinal fatty acid binding protein is also reportedly associated with LDLC levels [67]. Apolipoprotein B gene polymorphisms [68] and peroxisome proliferator activated receptor alpha polymorphism [69] reportedly influence TC and LDLC levels. Gene-gene interactions between these polymorphisms and Mt5178 C/A (ND2-237 Leu/Met) polymorphisms with regard to serum lipid levels must thus be considered.

Differences in the interaction between Mt5178 C/A (ND2-237 Leu/Met) polymorphism and habitual smoking with regard to serum lipid profiles were observed between subjects having BMI of $\leq 22$ and those having BMI of $> 22$. Changes in gene-environmental interactions on serum lipid levels with BMI have been reported [65]. Regrettably, we have no data that clarifies the reasons for these BMI-related differences. Control of body weight reportedly improves lipid profile [70]; however, whether body weight control and/or cessation of smoking improves TC or LDLC levels in current smokers with Mt5178A (ND2-237Met) warrants further longitudinal study.

Longevity-associated mitochondrial genotypic differences in the effects of BMI and habitual smoking on serum lipid levels may be the result of functional differences between the ND2-237 Leu/Met variants. This amino acid change may bring about physiological and/or biochemical changes in NADH dehydrogenase. Reactive oxygen species (ROS) are reportedly involved in atherogenesis [71] and aging [3], and the mitochondrial respiratory chain is an important source of ROS [72, 73]. Complex I of the mitochondrial respiratory chain (NADH dehydrogenase) consists of over forty different subunits [74] and plays a key role in the production of ROS [73, 75]. Cigarette smoking, which is known to cause increases in ROS [76], attenuates NADH dehydrogenase activity [77]. The mechanism by which ND2-237 Leu/Met polymorphism modulates the effects of cigarette smoking on intraocular pressure [24], serum protein fraction levels [26] and red blood cell counts [52] is also thought to result from physiological and biochemical differences between the ND2-237 Leu/Met variants. However, there is currently a lack of clear data and further investigation of these mechanisms is required.

A potential limitation of the present study was the evaluation of habitual drinking based on the frequency of alcohol consumption. The present study confirms the interaction between ND2-237 Leu/Met variants and habitual drinking with regard to TC and TG levels in subjects with BMI of ≤ 22. Although we also used this evaluation in previous reports [23, 24, 27], the question of whether the ND2-237 Leu/Met variants are sensitive to alcohol-induced effects on serum lipid levels in a dose-dependent manner warrants further investigation.

In conclusion, longevity-associated Mt5178 C/A (ND2-237 Leu/Met) polymorphism may influence the effects of habitual smoking on TG levels in Japanese men with BMI of ≤ 22, and on TC and LDLC levels in those with BMI of > 22. To the best of our knowledge, these are novel gene-environmental (lifestyle) interactions on TC and LDLC levels. Cigarette smoking and overweight may counteract the antiatherogenic effects of Mt5178A (ND2-237Met). These genetic epidemiological findings may contribute to the establishment of personalized prevention of lifestyle-related atherosclerotic diseases, which may be more effective than conventional prevention approaches.

## Acknowledgements

This study was supported in part by Grants-in-Aid from the Ministry of Education, Culture, Sports, Science and Technology of Japan (No. 14570355) and the Chiyoda Mutual Life Foundation.

## References

[1] Scheffler, I. E. (2001) A century of mitochondrial research: achievements and perspectives. *Mitochondrion 1*: 3-31.

[2] Van Remmen, H. & Richardson, A. (2001) Oxidative damage to mitochondria and aging. *Exp. Gerontol. 36*: 957-968.

[3] Driver, C. (2003) Mitochondrial interventions in aging and longevity. In S. I. S. Rattan (Ed.) *Biology of aging and its modulation vol. 5: Modulating aging and longevity*, (pp. 205-217). Dordrecht, The Netherlands: Kluwer Academic Publishers.

[4] Tanaka, M., Gong, J. S., Zhang, J., Yoneda, M. & Yagi, K. (1998) Mitochondrial genotype associated with longevity. *Lancet 351*: 185-186.

[5] Tanaka, M., Gong, J. S., Zhang, J., Yamada, Y., Borgeld, H. J. & Yagi, K. (2000) Mitochondrial genotype associated with longevity and its inhibitory effect on mutagenesis. *Mech. Ageing Dev. 116*: 65-76.

[6] Ivanova, R., Lepage, V., Charron, D. & Schächter, F. (1998) Mitochondrial genotype associated with French Caucasian centenarians. *Gerontology 44*: 349.

[7] De Benedictis, G., Rose, G., Carrieri, G., De Luca, M., Falcone, E., Passarino, G., Bonafé, M., Monti, D., Baggio, G., Bertolini, S., Mari, D., Mattace, R. & Franceschi, C. (1999) Mitochondrial DNA inherited variants are associated with successful aging and longevity in humans. *FASEB J. 13*: 1532-1536.

[8] Ross, O. A., McCormack, R., Curran, M. D., Duguid, R. A., Barnett, Y. A., Rea, I. M. & Middleton, D. (2001) Mitochondrial DNA polymorphism: its role in longevity of the Irish population. *Exp. Gerontol. 36*: 1161-1178.

[9] Niemi, A. K., Hervonen, A., Hurme, M., Karhunen, P. J., Jylhä, M. & Majamaa, K. (2003) Mitochondrial DNA polymorphisms associated with longevity in a Finnish population. *Hum. Genet. 112*: 29-33.

[10] Tanaka, M., Takeyasu, T., Fuku, N., Guo, L. J. & Kurata, M. (2004) Mitochondrial genome single nucleotide polymorphism and their phenotypes in the Japanese. *Ann. N.Y. Acad. Sci. 1011*: 7-20.

[11] Cann, R. L., Stoneking, M. & Wilson, A. C. (1987) Mitochondrial DNA and human evolution. *Nature 325*: 31-36.

[12] Yao, Y. G., Kong, Q. G. & Zhang, Y. P. (2002) Mitochondrial DNA 5178A polymorphism and longevity. *Hum. Genet. 111*: 462-463.

[13] Shimokata, H., Yamada, Y., Nakagawa, M., Okubo, R., Saido, T., Funakoshi, A., Miyasaka, K., Ohta, S., Tsujimoto, G., Tanaka, M., Ando, F. & Niino, N. (2000) Distribution of geriatric disease-related genotypes in the National Institute for Longevity Science, Longevity Study of Aging (NILS-LSA). *J. Epidemiol. 10*: S46-S55.

[14] Wang, D., Taniyama, M., Suzuki, Y., Katagiri, T. & Ban, Y. (2001) Association of the mitochondrial DNA 5178 A/C polymorphism with maternal inheritance and onset of type 2 diabetes in Japanese patients. *Exp. Clin. Endocrinol. Diabetes 109*: 361-364.

[15] Matsunaga, H., Tanaka, Y., Tanaka, M., Gong, J. S., Zhang, J., Nomiyama, T., Ogawa, O., Ogihara, T., Yamada, Y., Yagi, K. & Kawamori, R. (2001) Antiatherogenic mitochondrial genotype in patients with type 2 diabetes. *Diabetes Care 24*: 500-503.

[16] Mukae, S., Aoki, S., Itoh, S., Sato, R., Nishio, K., Iwata, T. & Katagiri, T. (2003) Mitochondrial 5178A/C genotype is associated with acute myocardial infarction. *Circ. J. 67*: 16-20.

[17] Ohkubo, R., Nakagawa, M., Ikeda, K., Kodama, T., Arimura, K., Akiba, S., Saito, M., Ookatsu, Y., Atsuchi, Y., Yamano, Y. & Osame, M. (2002) Cerebrovascular disorders and genetic polymorphisms: mitochondrial DNA5178C is predominant in cerebrovascular disorders. *J. Neurol. Sci. 198*: 31-35.

[18] Aoyama, M., Shidoji, Y., Saimei, M., Tsunawake, N. & Ichinose, M. (2003) Phenotypic linkage between single-nucleotide polymorphisms of $\beta_3$-adrenergic receptor gene and NADH dehydrogenase subunit-2 gene, with special reference to eating behavior. *Biochem. Biophys. Res. Commun. 309*: 261-265.

[19] Kato, T., Kunugi, H., Nanko, S. & Kato, N. (2000) Association of bipolar disorder with the 5178 polymorphism in mitochondrial DNA. *Am. J. Med. Genet. 96*: 182-186.

[20] Iwao N, Iwao S, Kobayashi F, Tajima K, Tanaka M, Atsuta Y, Tamakoshi A & Hamajima N. (2003) No association of the mitochondrial genotype (Mt5178A/C) with six cancers in a Japanese population. *Asian Pacific J. Cancer Prev. 4*: 331-336.

[21] Kokaze, A., Ishikawa, M., Matsunaga, N., Yoshida, M., Sekine, Y., Teruya, K., Takeda, N., Sumiya, Y., Uchida, Y. & Takashima, Y. (2001) Association of the mitochondrial DNA 5178 A/C polymorphism with serum lipid levels in the Japanese population. *Hum. Genet. 109*: 521-525.

[22] Takagi, K., Yamada, Y., Gong, J. S., Sone, T., Yokota, M & Tanaka, M. (2004) Association of a 5178C→A (Leu237Met) polymorphism in the mitochondrial DNA with a low prevalence of myocardial infarction in Japanese individuals. *Atherosclerosis 175*: 281-286.

[23] Kokaze, A., Ishikawa, M., Matsunaga, N., Yoshida, M., Sekine, Y., Sekiguchi, K., Harada, M., Satoh, M., Teruya, K., Takeda, N., Fukazawa, S., Uchida, Y. & Takashima, Y. (2004) Longevity-associated mitochondrial DNA 5178 A/C polymorphism and blood pressure in the Japanese population. *J. Hum. Hypertens. 18*: 41-45.

[24] Kokaze, A., Yoshida, M., Ishikawa, M., Matsunaga, N., Makita, R., Satoh, M., Sekiguchi, K., Masuda, Y., Uchida, Y. & Takashima, Y. (2004) Longevity-associated mitochondrial DNA 5178 A/C polymorphism is associated with intraocular pressure in Japanese men. *Clin. Experiment. Ophthalmol. 32*: 131-136.

[25] Kokaze, A., Ishikawa, M., Matsunaga, N., Yoshida, M., Sekine, Y., Teruya, K., Takeda, N., Satoh, M., Sumiya, Y., Uchida, Y. & Takashima, Y. (2002) Association of the longevity-associated mitochondrial DNA 5178 A/C polymorphism with serum protein fraction levels in healthy Japanese women. *Exp. Gerontol. 37*: 931-936.

[26] Kokaze, A., Ishikawa, M., Matsunaga, N., Yoshida, M., Sekine, Y., Sekiguchi, K., Satoh, M., Harada, M., Teruya, K., Takeda, N., Uchida, Y. & Takashima, Y. (2003) Longevity-associated mitochondrial DNA 5178 A/C polymorphism influences effects of cigarette smoking on serum protein fraction levels in Japanese men. *Mech. Ageing Dev. 124*: 765-770.

[27] Kokaze, K., Ishikawa, M., Matsunaga, N., Yoshida, M., Sekine, Y., Sekiguchi, K., Satoh, M., Harada, M., Teruya, K., Takeda, N., Uchida, Y., Tsunoda, T. & Takashima, Y. (2003) Longevity-associated mitochondrial DNA 5178 A/C polymorphism modulates effects of daily drinking and cigarette consumption on serum triglyceride levels in middle-aged Japanese men. *Exp. Gerontol. 38*: 1071-1076.

[28] Bailey, S. M., Pietsh, E. C. & Cunningham, C. C. (1999) Ethanol stimulates the production of reactive oxygen species at mitochondrial complex I and III. *Free Radic. Biol. Med. 27*: 891-900.

[29] Inoue, M., Sato, E. F., Park, A. M., Nishikawa, M., Kasahara, E., Miyoshi, M., Ochi, A. & Utsumi, K. (2001) Cross-talk between NO and oxyradicals, a supersystem that regulates energy metabolism and survival of animals. *Free Radic. Res. 33*: 757-770.

[30] Atlante, A., Seccia, T. M., Pierro, P., Vulpis, V., Marra, E., Pirrelli, A. & Passarella, S. (1998) ATP synthesis and export in heart left ventricle mitochondria from spontaneously hypertensive rat. *Int. J. Mol. Med. 1*: 709-716.

[31] Osanai, T., Tanaka, M., Kamada, T., Takahashi, K., Okada, S., Sirato, K., Magota, K., Kodama, S. & Okumura, K. (2001) Mitochondrial coupling factor 6 as a potent endogenous vasoconstrictor. *J. Clin. Invest. 108*: 1023-1030.

[32] Iwai, N., Katsuya, T., Mannami, T., Higaki, J., Ogihara, T., Kokame, K., Ogata, J. & Baba, S. (2002) Association between SAH, an acyl-CoA synthetase gene, and hypertriglyceridemia, obesity, and hypertension. *Circulation 105*: 41-47.

[33] Shiose Y. (1984) The aging effect on intraocular pressure in an apparently normal population. *Arch. Ophthalmol. 102*: 883-887.

[34] Klein, B. E. K., Klein, R. & Linton, K. L. P. (1992) Intraocular pressure in an American community. The Beaver Dam Eye Study. *Invest. Ophthalmol. Vis. Sci. 33*: 2224-2228.

[35] Bonomi, L., Marchini, G., Marraffa, M., Bernardi, P., De Franco, I., Perfetti, S., Varotto, A. & Tenna, V. (1998) Prevalence of glaucoma and intraocular pressure distribution in a defined population. The Egna-Neumarkt Study. *Ophthalmology 105*: 209-215.

[36] Lee, J. S., Lee, S. H., Oum, B. S., Chung, J. S., Cho, B. M. & Hong, J. W. (2002) Relationship between intraocular pressure and systemic health parameters in a Korean population. *Clin. Experiment. Ophthalmol. 30*: 237-241.

[37] Hayreh, S. S., Bill, A. & Sperber, G. O. (1994) Effects of high intraocular pressure on the glucose metabolism in retina and optic nerve in old atherosclerotic monkeys. *Graefe's Arch. Clin. Exp. Ophthalmol. 232*: 745-752.

[38] Harris, A., Jonescu-Cuypers, C., Martin, B., Kagemann, L., Zalish, M. & Garzozi, H. J. (2001) Simultaneous management of blood flow and IOP in glaucoma. *Acta Ophthalmol. Scand. 79*: 336-341.

[39] Chen, H. B., Yamabayashi, S., Ou, B. & Tsukahara S. (1996) Morphological changes in rabbit ciliary epithelium and blood-aqueous barriers after intravitreal 10 (-5) M endothelin-1. *Exp. Eye Res. 62*: 605-612.

[40] Ueda, J., Wentz-Hunter, K. K., Cheng, E. L., Fukuchi, T., Abe, H. & Yue, B. Y. J. T. (2000) Ultrastructural localization of myocilin in human trabecular meshwork cells and tissues. *J. Histochem. Cytochem. 48*: 1321-1329.

[41] Wentz-Hunter, K. K., Ueda, J., Shimizu, N. & Yue, B. Y. J. T. (2002) Myocilin is associated with mitochondria in human trabeculat meshwork cells. *J. Cell Physiol. 190*: 46-53.

[42] Herrmann, F. R., Safran, C., Levkoff, S. E. & Minaker, K. L. (1992) Serum albumin level on admission as a predictor of death, length of stay, and readmission. *Arch. Int. Med. 152*: 125-130.

[43] Sahyoun, N. R., Jacques, P. F., Dallal, G. & Russell, R. M. (1996) Use of albumin as a predictor of mortality in community-dwelling and institutionalized elderly populations. *J. Clin. Epidemiol. 49*: 981-988.

[44] Sullivan, D. H. (2001) What do the serum proteins tell us about our elderly patients? *J. Gerontol. 56A*: M71-M74.

[45] Wolf, G. T., Chretien, P. B., Weiss, J. F., Edwards, B. K. & Spiegel, H. E. (1982) Effects of smoking and age on serum levels of immune reactive proteins. *Otolaryngol. Head Neck Surg. 90*: 319-326.

[46] Banedek, I. H., Blouin, R. A. & McNamara, P. J. (1984) Influence of smoking on serum protein composition and the protein binding of drugs. *J. Pharm. Pharmacol. 36*: 214-216.

[47] Lindberg, G., Råstam, L., Gullberg, B., Lundbald, A., Nilsson-Ehle, P. & Hanson, B. S. (1993) Serum concentrations of total sialic acid and sialoglycoproteins in relation to coronary heart disease risk markers. *Atherosclerosis 103*: 123-129.

[48] Shima, M. & Adachi, M. (1996) Effects of environmental tobacco smoke on serum levels of acute phase proteins in schoolchildren. *Prev. Med. 25*: 617-624.

[49] Mori, T., Sasaki, J., Kawaguchi, H., Handa, K., Takada, Y., Matsunaga, A., Kono, S. & Arakawa, K. (1995) Serum glycoproteins and severity of coronary atherosclerosis. *Am. Heart J. 129*: 234-238.

[50] Hashimoto, S., Miwa, M., Akasofu, K. & Nishida, E. (1991) Changes in 40 serum proteins of post-menopausal women. *Maturitas 13*: 12-33.

[51] Denko, C. W. & Gabriel, P. (1981) Age and sex related levels of albumin, ceruloplasmin, alpha 1 antitrypsin, alpha 1 acid glycoprotein, and transferrin. *Ann. Clin. Lab. Sci. 11*: 63-68.

[52] Kokaze, A., Ishikawa, M., Matunaga, N., Yoshida, M., Makita, R., Satoh, M., Teruya, K., Sekiguchi, K., Masuda, Y., Harada, M., Uchida, Y. & Takashima Y. (2005) Interaction between longevity-associated mitochondrial DNA 5178 C/A polymorphism and cigarette smoking on hematological parameters in Japanese men. *Arch. Gerontol. Geriatr. 40*: 113-122.

[53] Tegos, T. J., Kalodiki, E., Sabeta, M. M. & Nicolaides, A. N. (2001) The genesis of atherosclerosis and risk factors: a review. *Angiology 52*: 89-98.

[54] Banzet, N., Francois, D. & Polla, B. S. (1999) Tobacco smoke induces mitochondrial depolarization along with cell death: effects of antioxidants. *Redox Rep. 4*: 229-236.

[55] Allain, C. C., Poon, L. S., Chan, C. S., Richmond, W. & Fu, P. C. (1974) Enzymatic determination of total serum cholesterol. *Clin. Chem. 20*: 470-475.

[56] Sugiuchi, H., Uji, Y., Okabe, H., Irie, T., Uekama, K., Kayahara, N. & Miyauchi, K. (1995) Direct measurement of high-density lipoprotein cholesterol in serum with polyethylene glycol-modified enzymes and sulfated alpha-cyclodextrin. *Clin. Chem. 41*: 717-723.

[57] Fossati, P. & Prencipe, L. (1982) Serum triglyceride determined colorimetrically with an enzyme that produces hydrogen peroxide. *Clin. Chem. 28*: 2077-2080.

[58] Friedewald, W. T., Levy, R. I. & Fredrickson, D. S. (1972) Estimation of the concentration of low-density lipoprotein cholesterol in plasma, without use of the preparative ultracentrifuge. *Clin. Chem. 18*: 499-502.

[59] Tokunaga, K., Matsuzawa, Y., Kotani, K., Keno, Y., Kobatake, T., Fujioka, S. & Tarui, S. (1991) Ideal body weight estimated from the body mass index with the lowest morbidity. *Int. J. Obes. 15*: 1-5.

[60] Uchigata, Y., Okada, T., Gong, J.S., Yamada, Y., Iwamoto, Y. & Tanaka, M. (2002) A mitochondrial genotype associated with the development of autoimmune-related type 1 diabetes. *Diabetes Care 25*: 2106.

[61] Nakanishi, N., Nakamura, K., Ichikawa, S., Suzuki, K. & Tatara, K. (1999) Relationship between lifestyle and serum lipid and lipoprotein levels in middle-aged Japanese men. *Eur. J. Epidemiol. 15*: 341-348.

[62] Umeda, T., Kono, S., Sakurai, Y., Shinchi, K., Imanishi, K., Nishikawa, H., Ogawa, S., Katsurada, M., Wakabayashi, K., Honjo, S. & Todoroki, I. (1998) Relationship of

cigarette smoking, alcohol use, recreational exercise and obesity with serum lipid atherogenicity: a study of self-defense officials in Japan. *J. Epidemiol. 8*: 227-234.

[63] Malin, R., Loimaala, A., Nenonen, A., Mercuri, M., Vuori, I., Pasanen, M., Oja, P., Bond, G., Koivula, T., Lehtimaki, T. (2001) Relationship between high-density lipoprotein paraoxonase gene M/L55 polymorphism and carotid atherosclerosis differs in smoking and nonsmoking men. *Metabolism 50*: 1095-1101.

[64] Corella, D., Guillen, M., Saiz, C., Portoles, O., Sabater, A., Folch, J. & Ordovas, J. M. (2002) Associations of LPL and APOC3 gene polymorphisms on plasma lipids in a Mediterranean population: interaction with tobacco smoking and the APOE locus. *J. Lipid Res. 43*: 416-427.

[65] Lussier-Cacan, S., Bolduc, A., Xhignesse, M., Niyonsenga, T. & Sing, C. F. (2002) Impact of alcohol intake on measures of lipid metabolism depends on context defined by gender, body mass index, cigarette smoking, and apolipoprotein E genotype. *Arterioscler. Thromb. Vasc. Biol. 22*: 824-831.

[66] Ikewaki, K., Mabuchi, H., Teramoto, T., Yamada, N., Oikawa, S., Sasaki, J., Taketa, K., Saito, Y. & Japan CETP Study Group. (2003) Association of cholesteryl ester transfer protein activity and TaqIB polymorphism with lipoprotein variation in Japanese subjects. *Metabolism 52*: 1564-1570.

[67] Galluzzi, J. R., Cupples, L. A., Otvos, J. D., Wilson, P. W. F, Schaefer, E. J. & Ordovas, J. M. (2001) Association of the A/T54 polymorphism in the intestinal fatty acid binding protein with variations in plasma lipids in The Framingham Offspring Study. *Atherosclerosis 159*: 417-424.

[68] Bentzen, J., Jørgensen, T. & Fenger, M. (2002) The effect of six polymorphisms in the apolipoprotein B gene on parameters of lipid metabolism in a Danish population. *Clin. Genet. 61*: 126-134.

[69] Tai, E. S., Demissie, S., Cupples, L. A., Corella, D., Wilson, P. W., Schaefer, E. J. & Ordovas, J. M. (2002) Association between the PPARA L162V polymorphism and plasma lipid levels. The Framingham Offspring Study. *Arterioscler. Thromb. Vasc. Biol. 22*: 805-810.

[70] Riebe, R., Greene, G. W., Ruggiero, L., Stillwell, K. M., Blissmer, B., Nigg, D. R. & Caldwell, M. (2003) Evaluation of a healthy-lifestyle approach to weight management. *Prev. Med. 36*: 45-54.

[71] Hogg N. (1998) Free radicals in disease. *Semin. Reprod. Endocrinol. 16*: 241-248.

[72] Wallace, D. C. (1992) Mitochondrial genetics: a paradigm for aging and degenerative diseases? *Science 256*: 628-632.

[73] Lenaz, G., Bovina, C., Aurelio, M. D'., Fato, R., Formiggini, G., Genova, M. L., Giuliano, G., Pich, M. M., Paolucci, U., Castelli, G. P. & Ventura, B. (2002) Role of Mitochondria in oxidative stress and aging. *Ann. NY Acad. Sci. 959*: 199-213.

[74] Walker, J. E., Skehel, J. M. & Buchanan, S. K. (1995) Structural analysis of NADH: ubiquinone oxidoreductase from bovine heart mitochondria. *Methods Enzymol. 260*: 14-34.

[75] Barja, G. (1998) Mitochondrial free radical production and aging in mammals and birds. *Ann. NY Acad. Sci. 854*: 224-238.

[76] Traber, M. G., van der Vliet, A., Reznick, A. Z. & Cross, C. E. (2000) Tobacco-related diseases. Is there role for anitioxidant micronutrient supplementation? *Clin. Chest Med. 21*: 173-187.

[77] Smith, P. R., Cooper, J. M., Govan, G. G., Harding, A. E. & Schapira, A. H. (1993) Smoking and mitochondrial function: a model for environmental toxins. *Q. J. Med. 86*: 657-660.

In: Trends in DNA Research
Editor: Corey R. Woods, pp. 53-64

ISBN 1-59454-635-5


*Chapter III*

# The Clinical Value of Static DNA Cytometry in the Diagnosis, Prognosis and Therapy of Prostate and Bladder Cancer

***F. St. Krause,* Th. Zoercher, K. M. Schrott and D. G. Engehausen***
Department of Urology, Friedrich-Alexander-University of Erlangen-Nuremberg, Germany

## Abstract

In order to identify prognostic factors regarding tumor diseases, which consequently determines the individual approach of therapy, it is necessary to first characterize the biological potential of the cancer. Standard methods in the diagnosis and prognosis of prostate and bladder carcinomas provide only limited information on the tumorbiological behavior. At present, the practical role of DNA cytophotometry is emerging. Apart from stage and tumor differentiation, it presents an independent criterion in the diagnosis of tumor rating. Furthermore, DNA cytophotometry helps to evaluate biological tumor progression and the consequent therapeutic approach. In carcinoma of the prostate we find it inevitable to determine the degree of DNA malignancy before therapy planning takes place. In addition, DNA cytometry evaluates hormone sensitivity of the tumor cells, especially in cases of heterogeinity of the prostate cancer. By using the stemline interpretation method of Fu we obtain information about the differential diagnosis and evaluation of the progressive character of prostate tumors. In regard to bladder carcinoma, the DNA image cytometry seems to be a prognostical tool in understanding

* Correspondence: Dr. Frens Steffen Krause, Department of Urology, Friedrich-Alexander-University of Erlangen-Nuremberg, Krankenhausstr. 12, 91054 Erlangen, Germany. phone: +49/9131/8223-868; fax: +49/9131/8223-179; email: frens.steffen.krause@t-online.de

the tumor biological behavior. Estimating of the risk of recurrence in patients after transurethral resection is possible by means of static DNA cytometry. Hence it may provide a valid criterion in the selection of adjuvant chemotherapy. But in the diagnosis of bladder cancer, the results and validity of regular clinical use of DNA image cytometry are controversial. For receiving the highest sensitivity of the measurement a combination of various DNA parameters is helpful, as we could demonstrate by our own scheme. Despite all criticisms and disadvantages, DNA image cytometry should be one of the tools of urological oncology, especially in diagnosing and for prognostic statements of prostate and bladder tumors.

## The Prostate Cancer

According to the statistics of Western societies, carcinoma of the prostate is the most frequent tumor of the male. This disease accounts for nearly 9% of all cancer deaths among men in USA and Europe [1]. The heterogeneity in prostate cancer is the reason for a difficult valuation for dignity and prognosis of this tumor. Therefore the discipline of urology must intensify efforts to treat prostate tumors. This carcinoma poses a problem because some cases proceed rather uneventfully over a period of years, even decades, whereas other cases present with a highly malignant course. Furthermore, a striking characteristic of prostate cancer is noted in its response to hormone therapy: 20% of prostate cancer cases do not primarily respond to hormones. Of the 80 % that initially respond positively, 50% show a progression of the disease again within 18 months, and 70% show progression within three years from diagnosis [2].

Regarding these aspects of the carcinoma's behavior, standard diagnostic procedures such as digital rectal examination, prostate specific antigen (PSA) testing, and transrectal ultrasound provide only limited information. However, studies by Tavares et al. [3, 4], Tribukait et al. [5, 6], and Boecking et al. [7, 8] on the value of DNA cytophotometry in the case of prostatic cancer articulate important statements about tumor biological behavior as noted above. Based on this information and our own examinations, the following value can be attached to DNA measurements in the presence of prostate carcinoma:

1. Besides assessing stage and malignancy of prostate carcinoma, DNA measurements serve as independent criteria in the diagnosis of tumor rating.
2. DNA readings provide a biological evaluation of tumor progression resulting in therapeutic consequences.
3. DNA measurements contribute information on hormone sensitivity of the prostate carcinoma.
4. The stemline interpretation of Fu is most efficient for diagnosis of prostate cancer, and allows prognostic statements of the disease.

These four essential statements are confirmed by literature: Studies by Tribukait [6] on the natural course of prostate cancer in untreated patients found a correlating grading and survival time. The author pointed out that survival time does not differ significantly in stages

A and B of prostate cancer, whereas stages C and D are associated with a considerable limitation of life expectancy. According to the results of Tribukait, tumor stage alone is not criteria enough to determine the survival chance in a patient. Also, histological and cytological grading do not suffice alone to pronounce a prognostic statement on the course of tumor disease. Presented for the first time by Tavares et al. [3], in 1966, the DNA cytometric malignancy versus grading evaluation of prostate cancer showed that life expectancy in patients presenting with stemline ploidy in 2c or in 4c after treatment with hormone therapy did not differ significantly from that of healthy men of the same age. However, survival rate of patients with stemlines in 3c or 6c dropped considerably. Such findings confirm the importance of DNA cytometric malignancy correlated to the grading in the evaluation of prostatic cancer.

While the tumor progresses, alterations in chromosomes of the tumor cell are detected by DNA malignancy grading of the prostate carcinoma. In 1987, Tribukait suggested a classification of the tumor by diploid, tetraploid, and aneuploid with one or several stemlines present, thus describing cytogenetic tumor progression [9]. Boecking et al. [10] suggests that in prostatic cancer, the cytogenetic step relevant to prognosis lies in the transition of a tetraploid to a hypo- and hypertetraploid, respectively, and aneuploid DNA distribution pattern.

In accordance with the currently applied therapeutic strategy in urological practice, implying a differentiation between locally defined and advanced prostate carcinoma it raises the question how relevant DNA cytometry is to the choice of therapy. Given a locally defined prostate carcinoma (T1+2, stage A+B), basic therapy possibilities include:

- total prostate vesiculectomy
- external radiation therapy and brachytherapy
- "wait and see".

A 1993 WHO consensus meeting in Stockholm passed the following statement [11], on the basis of study results by Boecking, Forsslund, and Tribukait. A strategy of "wait and see" is indicated only with diploid or tetraploid tumors. Since aneuploid tumors tend to respond badly to radiation and hormone therapy aggressive therapy such as total prostatectomy is the treatment of choice. However, controlled clinical therapy studies will be necessary to confirm the existing examination results. Examinations by Forsslund et al. [12] of hormone treated patients diagnosed with prostate cancer also differed, concerning prognostic statements, in diploid and/or tetraploid DNA distribution pattern as compared to aneuploid results.

In advanced prostate carcinoma (T3+4, stage A+B) hormone therapy plays an especially major role. Repeatedly, a quick clinical response from permanent androgen deprivation using LH-RH analogs does not last because androgen insensitivity develops with time. How can we account for this phenomenon? According to Tribukait [6] survival rate in patients with a diploid carcinoma of the prostate did not differ in hormone treated and untreated patients. In contrast, a difference was noted in tetraploid and aneuploid prostate cancer. Tribukait explains this phenomenon by the fact that advanced prostate carcinomas present with cell clones that are hormone dependent and or independent. Under hormone therapy, diploid tumors respond as hormone sensitive, thus identifying any hormone insensitive, aneuploid cell clones. When diploid cell clones are removed by hormone therapy, the remaining more aggressive, aneuploid cell clones determine the tumor event. Control biopsies performed because of unsuccessful hormone therapy confirm this postulation [13]. In our own experience so far, clinical tumor progression develops rather soon under hormone therapy when a patient presents from the start with not only diploid stemlines but also an additional, aneuploid stemline or even with a heterogenous aneuploid cell pattern. Further DNA examinations do not indicate a decrease in aneuploid clones. This finding is in accordance with the frequent clinical observation that many prostate carcinomas respond only initially to hormone therapy. Patients presenting with a tetraploid DNA distribution pattern are unsuitable for hormone therapy, because it stimulates a faster rate of tumor growth. This also applies to patients with an aneuploid DNA-distribution pattern when the tumor contains diploid or tetraploid elements.

The heterogeneity in prostate cancer is the reason for a difficult valuation for dignity and prognosis of this tumor. For that we looked for a correlation between PSA, tumor staging, and DNA cytophotometry [14]. 22 prostates (pT1-T4) from patients with prostate cancer who underwent radical prostatectomy were examined (table 1). Preoperative PSA and postoperative DNA image cytometry, after 2-8 needle biopsies out of each organ, were evaluated. We evaluated the DNA (counterstaining according to Feulgen [15]) from a minimum of 250 nuclei of diagnostic cells, calibtrated to 30 autologous lymphocytes (reference cells). For semiquantitative DNA examination we used the TV based image analysis system "Cires 3.1" (Zeiss / Kontron, Munich, Germany). The DNA interpretation was done to the single cell interpretation of Boecking [16] and stemline interpretations of Fu et al. [17] and the conventional one, with aneuploidy with a stemline > 2.2c. The prostate cancer tissues showed in DNA stemline interpretation of Fu if homogenous diploid an average PSA level of 3.8 ng/ml, and if homogenous aneuploid tumors a level of 14.0 ng/ml (table 2). Tumors with heterogeneous DNA patterns with a majority of aneuploidy had an average PSA level of 85.6 ng/ml, and heterogeneous tissues with a majority of diploidy a level of 10.9 ng/ml. We could not find a correlation for two of the three DNA interpretations (single cell interpretation of Boecking and conventional stemline interpretation) to the PSA or to the staging and grading systems (TMN and G, UICC 1992). Only the stemline interpretation of Fu after DNA cytophotometry showed a correlation between PSA and DNA ploidy in the tumor cells but no correlation to the histopathological findings. Our results are controversial to the results of Mora et al. [18] who described a correlation of DNA aneuploidy and high proliferative fraction with high Gleason grade and adverse prognosis but not with the PSA level in Stage B prostate cancer (pT2 N0 M0). Badalament et al. [19] demonstrated that DNA ploidy correlated significantly with pathologic stage; aneuploidy was identified more frequently in patients with stages C and D tumors. Significant racial differences were found in that study with respect to DNA ploidy, mean DNA indices, and mean PSA values. Our data also demonstrated a typical problem with different models of DNA cytophotometry; the results differ in a bigger range for the exact DNA ploidy status. For example a patient with pT3a N0 G2 prostate cancer with a serum PSA of 10.1 ng/ml shows in single cell interpretation of Boecking a homogenous diploid status, in stemline interpretation of Fu a heterogeneous status with a majority of diploidy and in conventional stemline interpretation a homogenous aneuploid status, so the evaluation of the progressive character of the prostate cancer from this patient is not possible. Jones et al. presented in aneuploid tissues of prostate cancer sensitivity in DNA cytometry from 50% to 61% [20].

**Table 1. Description of prostate cancer cases (n = 22) with staging (TNM), grading (histology) and PSA: 2-8 DNA measurements per case evaluated according to three different single cell (SCI) and stemline interpretations (SLI)**

| no | TNM | hist. | PSA (ng/ml) | DNA cytophotometry | | | | | |
|---|---|---|---|---|---|---|---|---|---|
| | | | | SCI-Boecking | | SLI-FU | | SLI-conv. | |
| | | | | diploid | aneupl | diploid | aneupl | diploid | Aneupl |
| 1 | T1N0MX | G1 | 1.9 | 2 | 0 | 1 | 1 | 2 | 0 |
| 2 | T1N0MX | G1 | 1.4 | 5 | 0 | 3 | 2 | 5 | 0 |
| 3 | T2N0MX | G2 | 1.4 | 3 | 0 | 2 | 1 | 2 | 1 |
| 4 | T2N0MX | G1 | 7.5 | 4 | 0 | 2 | 2 | 3 | 1 |
| 5 | T2N0MX | G2 | 2.8 | 2 | 0 | 2 | 0 | 2 | 0 |
| 6 | T2N0MX | G2 | 7.5 | 7 | 0 | 5 | 2 | 5 | 2 |
| 7 | T2N0MX | G2 | 1.1 | 2 | 0 | 2 | 0 | 2 | 0 |
| 8 | T2N0MX | G1 | 10.9 | 5 | 0 | 3 | 2 | 4 | 1 |
| 9 | T3N1MX | G3 | 161.0 | 5 | 3 | 5 | 3 | 8 | 0 |
| 10 | T3N0MX | G3 | 11.8 | 2 | 0 | 0 | 2 | 0 | 2 |
| 11 | T3N0MX | G2 | 35.7 | 6 | 0 | 5 | 1 | 6 | 0 |
| 12 | T3N0MX | G2 | 10.1 | 6 | 0 | 4 | 2 | 0 | 6 |
| 13 | T3N0MX | G2 | 9.3 | 2 | 0 | 2 | 0 | 2 | 0 |
| 14 | T3N1MX | G2 | 16.2 | 1 | 1 | 0 | 2 | 1 | 1 |
| 15 | T3N0MX | G2 | 3.3 | 4 | 0 | 4 | 0 | 4 | 0 |
| 16 | T3N0MX | G2 | 3.0 | 2 | 0 | 2 | 0 | 2 | 0 |
| 17 | T3N0Mx | G3 | 8.1 | 5 | 3 | 6 | 2 | 8 | 0 |
| 18 | T3N0MX | G2 | 14.1 | 6 | 1 | 5 | 2 | 5 | 2 |
| 19 | T3N0MX | G2 | 5.2 | 2 | 0 | 1 | 1 | 1 | 1 |
| 20 | T3N0MX | G2 | 9.7 | 5 | 0 | 4 | 1 | 4 | 1 |
| 21 | T4N0MX | G3 | 10.2 | 6 | 0 | 2 | 4 | 5 | 1 |
| 22 | T4N0MX | G3 | 3.0 | 5 | 0 | 5 | 0 | 5 | 0 |

**Table 2. DNA stemline interpretation of Fu of 22 prostate cancer cases as homogenous and heterogenous DNA patterns correlated to PSA values (mean and range)**

| | homogenous DNA | | heterogeneous DNA | | |
|---|---|---|---|---|---|
| | Diploid | aneuploid | dipl>aneupl | dipl=aneupl | dipl<aneupl |
| n = 22 | 6 | 2 | 9 | 3 | 2 |
| PSA mean (ng/ml) | 3.8 | 14.0 | 10.9 | 4.9 | 85.6 |
| PSA range (ng/ml) | 1.1-9.3 | 11.8-16.2 | 1.4-35.7 | 1.9-7.5 | 10.2-161.0 |

In spite of these different aspects on DNA ploidy in patients with diagnosed prostate cancer, it becomes inevitable to determine the DNA malignancy degree before selecting a definite therapy, and to possibly address the issue of life expectancy in the carcinoma patient, especially the untreated one! For a diploid carcinoma of the prostate, remaining in the diploid state, is not likely to shorten the life span. Studies by Tavares et al. [3, 4] and Tribukait [5, 6] support this view. According to Boecking [21], the question arises to which extent certain therapies affect morbidity and mortality in a patient. This applies in particular to whether or not to use hormone therapy for prostate carcinoma because of its possible side effects.

## The Bladder Cancer

Urinary bladder carcinoma is considered the most frequent, malignant tumor of the urinary tract, and presents to large extend (93%) as transitional cell carcinoma of the bladder (TCC). In both sexes incidence is increasing; in 3.5% of cancer patients it is the cause of death. Bladder carcinomas are clinically defined as superficial (Ta, T1), muscle invasive (T2), and organ overlapping (T3, T4), and by histopathological grading as well (G1), moderate (G2), and poorly differentiated (G3) [22]. Superficial and muscle invasive tumors are treated operatively by differentiated transurethral bladder resection [23]. Adjuvant chemotherapy is indicated in the presence of multiple bladder foci. The actual difficulty in the treatment of superficial bladder carcinomas lies in their frequent recurrence, at rates up to 70 % [24].

In order to identify prognostic factors regarding TCC, which consequently determines the individual approach of therapy, it is necessary to first characterize the biological potential of the tumor. It is not easy to predict recurrent disease of the tumor, because standard procedures like cytoscopy, cytology and different tested tumor markers (NMP22, BTA test, p53, etc.) have yielded unsatisfactory results. Therefore DNA cytophotometry may prove to be a more useful method, because it is more sensitive [25]. DNA cytophotometric studies by Tavares et al. [3], Taubert et al. [26, 27], and Al-Abadi et al. [28] provide valid information on the recurrence and progression of bladder cancer.

On the basis of these studies as well as our own findings, in urinary bladder cancer the subsequent significance may be attributed to DNA analysis:

1. Next to tumor stage and differentiation, DNA analysis functions as an independent tool for pathological diagnostics.
2. DNA analysis permits an evaluation of tumor biology and consequently an estimate of its aggressive potential as required for prognosis.
3. It could be useful to combinate various DNA parameters to increase the sensitivity of DNA examination.
4. DNA examination serves as additional criterion when deciding on adjuvant chemotherapy after tumor resection.

Examinations by Fossa et al. [29], Koss et al. [30], and Al-Abadi et al. [28] agree on a positive correlation of DNA ploidy with the histopathological grade and the tumor stage. Highly differentiated bladder carcinomas are characterized by diploid DNA patterns, whereas an aneuploid DNA distribution is distinctive of low differentiation, invasive tumors and carcinoma in situ. In a first prospective study by Al-Abadi et al. [28], neither metastasis nor tumor progression occurred in 127 patients (pT1-4, G1-3) with diploid tumors during nine years of observation. On the other hand, recurrent tumors and local tumor growth observed during 6 to 36 months were related to aneuploid cell lines. In comparison to grade and stage, DNA ploidy maintained the highest prognostic value. Analyzing the progress of DNA ploidy in 12 patients after transurethral resection of bladder carcinoma, we found that recurring tumors were associated either with a proliferating tetraploid stemline (4c) or with a diploid stemline (2c) with additional tetraploid cell clones present. Such tetraploid cell clones occurring during tumor progression were closely linked with recurrent disease.

Under clinical aspects for the use of DNA image cytometry in detection of bladder cancer we evaluated 151 bladder washings from patients suspected of bladder cancer [31] before transurethral resection. The pathologist diagnosed Ta-T4 transitional cell carcinomas next to negative findings. In this study we examined the clinical value of various common methods of DNA single cell and stemline interpretations. We evaluated the DNA (counterstaining according to Feulgen [15]) from a minimum of 250 nuclei of diagnostic cells, calibtrated to 30 autologous lymphocytes (reference cells). For semiquantitative DNA examination we used the TV based image analysis system "Cires 3.1" (Zeiss / Kontron, Munich, Germany). Comparing the specificity and sensitivity of DNA image cytometry in detection of bladder tumors, we found 81% and 52%, respectively, for single cell interpretation of Boecking [16], 84% and 45% for stemline interpretation of Boecking et al. [32], 61% and 58% for stemline interpretation of Fu et al. [17], and 82% and 40% for conventional stemline interpretation. In the attempt to improve DNA image cytometry, we subdivided DNA measurement into its single parameters and evaluated them according to our own scheme. A subdivision to the DNA-histogram may be necessary in order to facilitate interpretation, thus avoiding to be misled by false aneuploidy [33]. In consequence, we identified six single DNA parameters: DNA malignancy grade (DNA-MG), diploid deviation quotient (DDQ), mean ploidy (MP), entropy, 5c exceeding rate (5cEE) and aneuploid events diploid (AED). Out of all recorded measurements these parameters correlated most to histopathological grading (table 3). Creating reference values at random and rating by points, we used a cytometric grading system for ranking (tables 4 and 5). In our study we were able to increase sensitivity and specificity of DNA image cytometry to almost 70% in detecting TCC (table 6). In pursuit of the same purpose of improving the method of DNA cytometry, Van Velthoven et al. [34] evaluated nuclear DNA content, nuclear morphometry and chromatin patterns; Milord [35] measured different nuclear features (e.g. area, diameter, roundness, ellipticity) as part of his image analysis system. Nevertheless, both studies obtained no clinical benefit. Other possibilities for correcting valence of DNA cytometry, is to employ laser scanning cytometry [36] or to combine DNA flow cytometry with antigen testing such as for cytokeratins [37]. However, for routine diagnostics of bladder cancer both methods are too costly and require special lab equipment. Nowadays, modern methods of proving genetic alterations such as fluorescence in situ hybridisation (FISH), are used in the laboratories. Their published results on detection of transitional cell carcinoma of the bladder by FISH have been excellent compared to those of DNA cytometry. Their disadvantages, however, are the manipulation on a large scale and expensiveness of these methods, which altogether make them not easy to use routinely in the clinical setting [38, 39].

**Table 3. Median measured values of DNA parameters in correlation to histopathological findings (G1-G3), (*p<0.05, **p<0.005, ***p<0.001) of bladder tumors**

| DNA histo. | DNA-MG | DDQ | MP | Entropy | 5cER | AED |
|---|---|---|---|---|---|---|
| neg. | 0.33 | 1.11 | 2.25 | 3.18 | 2.26 | 4.02 |
| G1 | 0.46** | 1.14 | 2.27 | 3.48* | 2.35 | 5.34 |
| G2 | 0.97*** | 1.45** | 2.89** | 4.02** | 11.49** | 22.71** |
| G3 | 1.01 | 1.56 | 3.12 | 4.04 | 11.83 | 25.18 |

**Table 4. Determination of reference range for measured mean values of the six DNA parameters and rating by points (0-3)**

| DNA pts. | DNA-MG | DDQ | MP | Entropy | 5cER | AED |
|---|---|---|---|---|---|---|
| 0 | <0.35 | <1.1 | <2.2 | <3.1 | <2.3 | <5.0 |
| 1 | 0.35-0.65 | 1.1-1.3 | 2.2-2.6 | 3.1-3.55 | 2.3-7.0 | 5.0-13.0 |
| 2 | 0.66-1.0 | 1.31-1.5 | 2.61-3.0 | 3.56-4.0 | 7.0-11.5 | 14.0-24.0 |
| 3 | >1.0 | >1.5 | >3.0 | >4.0 | >11.5 | >24.0 |

**Table 5. Conversion of the total amount of points ($\sum$) into cytometrical grading (G1-G3)**

| $\sum$ points | cytometric grading |
|---|---|
| 0 – 2 | neg. |
| 3 – 7 | G1 |
| 8 – 13 | G2 |
| 14 – 18 | G3 |

**Table 6: Validity (specificity and sensitivity) of various single cell and stemline DNA interpretations besides our designed DNA interpretation scheme, at bladder cancer patients**

| DNA value | 5cEE (SCI) | Böcking (SLI) | FU (SLI) | conv. (SLI) | own scheme |
|---|---|---|---|---|---|
| specificity | 80.6% | 83.6% | 61.2% | 82.1% | 69.1% |
| sensitivity | 52.4% | 45.2% | 58.3% | 40.4% | 71.6% |

Proliferating tetraploid cells correspond tumorbiologically to the 5c exceeding rate, a criterion applied by Taubert et al. [26, 27] to predict tumor recurrences. Thus the prognostic value of various DNA parameters was examined in 75 patients diagnosed with urothelial cancer of the bladder. In the measured DNA parameters "aneuploidy of DNA stemline", "5cER > 5%", and "DNA malignancy degree > 0.5", the tumors differed quite significantly. Consequently, a DNA cytophotometric risk classification with these parameters was designed to provide a basis for therapy planning. If there is an indication of at least one of the parameters noted above, patients are advised to undergo adjuvant, topical chemotherapy that otherwise would not be required. The essential conclusion of this project is that DNA cytometry alone, and no other grading or staging parameter, was able to predict a non-recurrence of Ta, T1, and T2 urothelial carcinoma after resection, and that with a specificity of more than 90%. Therapy is more efficiently planned when this risk classification is taken into account. A study by Scheida [40] also back up such findings. He gave proof that TaG1 tumors, as well as other superficial tumors with a normal DNA pattern, do not benefit from prophylactic topical chemotherapy. In reference to his clinical background, Turkolmez et al. [41] described using DNA ploidy to identify patients likely to respond positively to therapy and extended life duration, after administration of neoadjuvant chemotherapy in treatment of invasive bladder cancer. Longatto et al. [42] and Desgrippes et al. [43] described clinical advantages of DNA cytometry when they diagnosed the majority of recurrent bladder tumors by combining cytological examination with DNA cytometric evaluation.

In conclusion we find that the examination of DNA ploidy in patients with bladder cancer is of highest prognostic value and opens up new possibilities to plan therapy efficiently. So, regardless of tumor grade and stage, DNA malignancy grading helps to identify patient groups that do not require adjuvant topical chemotherapy due to a low risk of recurrent disease. We were able to show that sensitivity of DNA examination can be increased by combining various DNA parameters. Apart from our own scheme, the discrepancy in interpretation of DNA image cytometry does not allow us to recommend this procedure as the only diagnostic in detection of bladder cancer. However, in regard to prognostic statements, particularly tumor biological behavior, DNA image cytometry appears to be useful.

## References


[1] Landis, SH; Murray, T; Bolden, S; Wingo, PA (1998). Cancer statistics. *Cancer J Clin* 48, 6-29.

[2] Flüchter SH. Das Restaging zur Bewertung der Therapie der ersten Wahl. In: Bichler KH, editor. *Prostatakarzinom*. Frankfurt: progammed Verlag; 1996; 30-34.

[3] Tavares, AS; Costa, J; de Carvalho, A; Reis, M (1966). Tumor ploidy and prognosis in carcinomas of the bladder and prostate. *Br J Cancer* 20, 438-441.

[4] Tavares, AS; Costa, J; Costa, M (1973). Correlation between ploidy and prognosis in prostatic carcinoma. *J Urol* 109, 676-679.

[5] Tribukait, B (1991). DNA flow cytometry in carcinoma of the prostate for diagnosis, prognosis and study of tumor biology. *Acta Oncol* 30, 187-192.

[6] Tribukait, B (1993). Nuclear deoxyribonucleic acid determination in patients with prostate carcinomas: Clinical research and application. *Eur Urol Su*ppl 2, 64-76.

[7] Boecking, A; Chatelain, R; Orthen, U; Gien G; von Kalkreuth, G; Jocham, D; Wohltmann, D (1988). DNA grading of prostatic carcinoma: Prognostic validity and reproducibility. *Anticancer* Res 8, 129-136.

[8] Boecking, A. Diagnostic DNA cytometry of prostatic cancer. Symposium on prostate cancer (part II) (1992). *Diagn Oncol* 2, 90-102.

[9] Tribukait, B (1987). Flow cytometry in assessing the clinical aggressiveness of genito urinary neoplasms. *World J Urol* 5, 108-122.

[10] Boecking, A; Striepecke, E; Auer, H; Fuezesi, L. Static DNA cytometry. Biological background, technique and diagnostic interpretation. In: Wied GL, Barthels PH, Rosenthal DL, Schenk U editors. Compendium on the computerized cytology and histology laboratory. Chicago: *Tutorials of Cytology;* 1994; 325-341.

[11] Schroeder, F; Boecking, A; de Verre White, R; Koss, L; Lieber, M; Stenkvist, B; Zetterberg, A (1994). Clinical utility of cellular DNA measurements in prostate carcinoma. *Scand J Urol Neprhrol Suppl* 162, 51-64.

[12] Forsslund, F; Esposti, P; Nilsson, B; Zetterberg, A (1992). The prognostic significance of nuclear DNA content in prostatic carcinoma. *Cancer* 69, 1432-1439.

[13] Feil, G; Mittermueller, B; Nelde, HJ; Wunderer, A; Wechsel, HW; Bichler KH. DNA-Bildzytometrie beim Prostatakarzinom. In: Bichler KH, editor. *Prostatakarzinom.* Frankfurt: progammed Verlag; 1996; 17-20.

[14] Krause, FS; Feil, G; Bichler, KH; Schrott, KM; Akcetin, ZY; Engehausen, DG (2004). Heterogenity in prostate cancer: Prostatic specific antigen (PSA) and DNA cytophotometry. *Anticancer Res* (accepted for publication).

[15] Cremer, TH; Jauch, A; Ried, TH; Schoeck, E; Lengauer, CH; Cremer, M; Speicher, MR (1995). Fluoreszenz in situ Hybridisierung. *Dt Aerzteblatt* 22, 1177-1185.

[16] Boecking, A (1990). DNA Zytometrie und Automation in der klinischen Diagnostik. *Beitr Onkol* 38, 298-347.

[17] Fu, YS; Cheng, L; Huang, I; Huang, S; Wiesmeier, E; Wettstein, F; Weissman, M (1989). DNA ploidy analysis of cervical condyloma and intraepithelial neoplasia in specimens obtained by punch biopsy. *Anal Quant Cytol Histol* 11, 187-195.

[18] Mora, LB; Moscinski, LC; Diaz, JI; Blair, P; Cantor, AB; Pow Sang, JM (1999). Stage B prostate cancer: Correlation of DNA ploidy analysis with histological and clinical parameters. *Cancer Control* 6, 587-591.

[19] Badalament, RA; O'Toole, RV; Young, DC; Drago, JR (1991). DNA ploidy and prostate specific antigen as prognostic factors in clinically resectable prostate cancer. *Cancer* 67, 3014-3023.

[20] Jones, EC; McNeal, J; Bruchowsky, N; de Jong, G (1990). DNA content in prostatic adenocarcinoma. A flow cytometric study of the predictive value of aneuploidy for tumor volume, percentage Gleason grade 4 and 5, and lymph node metastases. *Cancer* 66, 752-757.

[21] Boecking, A. Das lokal begrenzte Prostatakarzinom. Kriterien zur Aggressivität. In: Bichler KH, editor. *Prostatakarzinom.* Frankfurt: progammed Verlag; 1996; 24-25.

[22] Fischer, CG; Waechter, W; Fuentecilla Perez, E; Miller, J; Weidner, W; Dudeck, J (1997). Urologische Tumoren in Deutschland. *Urologe A* 36, 143-150.

[23] Bichler, KH; Wechsel, HW; Lahme, S; Wilbert, DM; Strohmaier, WL. Die Stellung der transurethralen Elektroresektion in der Behandlung des Harnblasenkarzinoms (differenzierende transurethrale Resektion). In: Bichler, KH, editor. *Diagnostik und Therapie des Harnblasenkarzinoms.* Reinbek: Einhorn Presse Verlag; 1998; 82-91.

[24] Otto, U; Huland, H; Klosterhalfen H. Rezidivprophylaxe oberflächlicher Blasenkarzinome. In: Huland H, Klosterhalfen H editors. Therapie und

Rezidivprophylaxe oberflächlicher *Harnblasenkarzinome.* Stuttgart: Thieme Verlag; 1984.

[25] Borchers, H; Planz, B; Jakse, G; Boecking, A (1994). DNA Aneuploidy in G1 Urothelial Carcinomas of the Urinary Bladder. *Urol Int* 52, 145-153.

[26] Taubert, G; Rassler, J; Nenning, H; Rose, S (1993). Die Bedeutung zytometrischer Parameter bei der Therapieplanung von Urothelkarzinomen. *Verh Dtsch Ges Pathol* 77, 216-221.

[27] Taubert, G; Nenning, H; Rassler, J; Scheida, U (1995). Prognostischer Wert von DNA Parametern bei oberflächlichen Urothelkarzinomen. *Verh Dtsch Ges Pathol* 79, 584.

[28] Al-Abadi, H; Nagel, R (1994). Deoxyribonucleic acid content and survival rates of patients with transitional cell carcinoma of the bladder. *J Urol* 151, 37-42.

[29] Fossa, SD; Kaalhus, O; Scott-Knudsen O (1977). The clinical and histopathological significance of Feulgen DNA values in transitional cell carcinoma of the human urinary bladder. *Europ J Cancer* 12, 1155-1162.

[30] Koss, LG ; Wersto, RP ; Simmons, DA ; Deitch, D ; Herz, F ; Freed, SZ (1989). Predictive value of DNA measurements in bladder washings. Comparision of flow cytometry, image cytophotometry and cytology in patients with a past history of urothelial tumors. *Cancer* 64, 916-924.

[31] Krause, FS; Feil, G; Bichler, KH; Schrott, KM; Akcetin, ZA (2003). Clinical aspects for the use of DNA image cytometry in detection of bladder cancer: A valuable tool?. DNA and Cell Biology 22, 721-725.

[32] Boecking, A; Biesterfeld, S; Liu, SQ. DNA distribution in gastric cancer and dysplasia. In: Zhang YS, editor. Precancerous conditions and lesions of the stomach. Heidelberg: *Springer Verlag;* 1993; 103-120.

[33] Bellaoui, H; Chefchaouni, MC; Lazrak, N; Khalfaoui, LC; Yassine, F; Elhamany, Z (2002). Flow cytometric DNA analysis and cytology in diagnosis and prognosis of bladder tumors. *Ann Urol* 36, 45-52.

[34] Van Velthoven, R; Petein, M; Oosterlinck, W; de Wilde, W; Mattelaer, J; Hardeman, M; Kiss, R; Decaestecker, C (2000). Identification by quantitative chromatin pattern analysis of patients at risk for recurrence of superficial transitional bladder carcinoma. *J Urol* 146, 2134-2137.

[35] Milord, RA; Lecksell, K; Epstein, JL (2001). An objective parameter to aid in the diagnosis of flat urothelial carcinoma in situ. *Hum Pathol* 32, 997-1002.

[36] Kawamura, K; Tanaka, T; Ikeda, R; Fujikawa Yamamoto, K; Suzuki, K (2000). DNA ploidy analysis of urinary tract epithelial tumors by laser scanning cytometry. *Anal Quant Cytol Histol* 22, 26-30.

[37] Sanchez Carbayo, M; Ciudad, J; Urrutia, M; Navajo, JA; Orfao, A (2001). Diagnostic performance of the urinary bladder carcinoma antigen ELISA test and multiparametric DNA / cytokeratin flow cytometry in urine voided samples from patients with bladder carcinoma. *Cancer* 92, 2811-2819.

[38] Cianciulli, AM; Bovani, R; Loenardo, C; Iori, F; Coletta, AM; Marzano, R; Antenucci, A; Gandolfo, GM; Laurenti, C (2001). DNA aberrations in urinary bladder cancer detected by flow cytometry and FISH: prognostic implications. *Eur J Histochem*, 65-71.

[39] Krause, FS; Feil, G; Zumbragel, A; Bichler, KH (2000). Molecular genetic methods in the diagnosis of invasive bladder cancer. *Anticancer Res* 20, 5015-5021.

[40] Scheida, U (1994). DNA zytometrische Untersuchungen an oberflächlichen Harnblasenkarzinomen (retrospektive Studie an paraffineingebetteten Material). *Diss. med. Leipzig.*

[41] Turkolmez, K; Baltaci, S; Beduk, Y; Muftuoglu, YZ; Gogus, O (2002). DNA ploidy and S phase fraction as predictive factors of response and outcome following noeadjuvant MVEC chemotherapy for invasive bladder cancer. *Scand J Urol Nephrol* 36, 46-51.

[42] Longatto Filho, A; Santinelli, A; Montironi, R (2001). Cytometric investigations of bladder irrigation / washings and voided urine. *Pathologica* 93, 164-167.

[43] Desgrippes, A; Izadafir, V; Assailly, J; Fontaine, E; Beurton, D (2000). Diagnosis and prediction of recurrence in superficial bladder cancers with DNA image cytometry and urinary cytology. *BJU Int* 85, 434-435.

In: Trends in DNA Research
Editor: Corey R. Woods, pp. 65-102
ISBN 1-59454-635-5


*Chapter IV*

# Molecular Mechanisms of Interaction of THC-apoA-I Complex with DNA and Initiation of Transcription

*L. E. Panin*[*]
Institute of Biochemistry, Siberian Branch of Russian Academy of Medical Sciences, Novosibirsk, Russia

## Abstract

It was shown that the cortisol-apolipoprotein A-I complex does not affect the rate of DNA and protein biosynthesis in primary culture of hepatocytes, while the tetrahydrocortisol-apolipoprotein A-I complex (THC-apoA-I) considerably accelerates the insertion of 3H-thymidine into DNA and 14C-leucine into protein. Molecular mechanism of this phenomenon was studied.

The in vitro action of the apolipoprotein A-I-tetrahydrocortisol complex on a native DNA isolated from mature rat liver was shown to cause the formation of S1-nuclease-sensitive regions. The most interactive sites were distributed over the genome irregularly. After the interaction with tetrahydrocortisol-apolipoprotein A-I complex, DNA gains the capacity of dose-dependent and highly cooperative binding to RNA-polymerase. The reduced form of hormone, tetrahydrocortisol, that was previously considered to be an inactive metabolite, when being composed with the complex with apolipoprotein A-I gains ability to carry out the biological function similar to the function of a transcriptional factor. Using the method of small-angle X-ray scattering (SAXS) it was demonstrated that a pronounced interaction with eukaryotic DNA is observed only in the case of THC-apoA-I complex. This interaction is accompanied by local melting of DNA. The most probable site of binding this complex to DNA is a sequence of (GCC)n type composing various eukaryotic genes, human genes as well. Oligonucleotide (duplex) of

[*] Corresponding author. Tel.: (3832) 322735; Fax: (3832) 336758; email: ibch@soramn.ru

this type has been synthesized. Its interaction with THC-apoA-I complex was shown to cause the formation of a more composite complex, which dissociates into complementary oligonucleotides. The latter ones also interact with THC-apoA-I complex. The kinetics of this multistep process is presented. Interaction of the cortisol-apolipoprotein A-I complex with the duplex is less specific and does not lead to the formation of complementary oligonucleotides.

Study of the mechanism of interaction between duplex CC(GCC)5 · GG(CGG)5Li2 and THC or THC-apoA-I complex revealed the formation of hydrogen bonds between OH group of the THC A-ring and C=O group of cytosine or guanine. Interaction with cortisol or cortisol-apoA-I complex leads to the formation of hydrogen bonds with NH group of cytosine. In addition, THC and cortisol form hydrogen bonds with PO2 group of the duplex and with OH group of monosaccharide. Interaction of apoA-I with the duplex is accompanied by the formation of hydrogen bonds with NH2 group of protein, with C=O group of cytosine, and P=O group of desoxyribose phosphate. Interaction of THC, apoA-I and THC-apoA-I complex with the duplex occurs at the same active site, namely, with C=O group. The order → disorder structural transition takes place in the duplex under the action of apoA-I or THC-apoA-I and cortisol-apoA-I complexes.

Thus, it was shown for the first time that reduced forms of steroid hormones have a significant biological effect. In complex with apolipoprotein A-I they enhance the rate of protein biosynthesis in parenchymal cells. The reported mechanism underlies intracellular regeneration.

## Introduction

An action of HDL on chromosomal DNA from different tissues and cells attracts considerable attention of researchers. At present, the mitogen-stimulating function of HDL is widely discussed in the literature. The stimulating effect of HDL was demonstrated for the proliferation of lymphocytes [1], epithelial [2], endothelial [3], smooth muscle [4], and tumor [5] cells. According to Desanctis et al. [6] all classes of lipoproteins enhanced proliferation of T cells and natural killer cells. However, HDL (unlike all other classes of lipoproteins) has no influence on interleukin-2-induced T cell proliferation, but enhanced proliferation of large granular lymphocytes 2-fold. In another work, these authors demonstrated that HDL decreased secretion of certain cytokines from the natural killer cells, including TNF-$\alpha$, granulocyte/macrophage colony-stimulating factor, and interleukin-1 and interleukin 1$\beta$. Secretion of interleukin-2, interleukin-8, and interferon-$\gamma$ was stimulated [7]. HDL inhibited cytokine-induced (tumor necrosis factor-$\alpha$; TNF-$\alpha$) expression of adhesion factor in human umbilical vein endothelial cells. Interestingly, $HDL_3$ fraction was more potent than $HDL_2$ fraction [8].

Apolipoprotein A-I (apoA-I) constitutes about 70% of the total amount of human high-density lipoproteins, which indicates its significant structural role. The function of apoA-I in the organism is still not clearly understood. ApoA-I is known to increase the activity of lecithin-cholesterol-acyltransferase [9]. The amphipathic alphahelical segments of molecules are responsible for the manifestation of this property. It is known also that the incubation of human trophoblast cells with apoA-I increases the placental lactogen expression, while the presence of actinomycin D completely inhibits the lactogen synthesis [10]. Apo A-I inhibited the IgG-induced activation of neutrophils. The addition of apoA-I to activated neutrophils decreased the release of granules and production of superoxide anion radicals by 60% [11].

Lipoproteins are known to serve as the most important transport form of steroid hormones in blood [12], with apolipoprotein A-I being the protein responsible for HDL binding and transport [13]. This is very important, since this protein was found in the fraction of acid nonhistone proteins of various type cells, hepatocytes among them [14]. Its content in transcriptionally active chromatin and nuclear matrix was the highest one [15].

Previously it was shown that glucocorticoids and high-density lipoproteins exert a cooperative effect, activating biosynthesis of RNA and a protein in the white rat liver [16, 17, 18]. The mechanism of this phenomenon differs from the known mechanism of the glucocorticoid induction of the gluconeogenesis enzymes [13]. Further studies revealed that the active form of hormone, acting in nucleus in concert with apoA-I, is its reduced form – tetrahydrocortisol [19], which was earlier considered as an inactive metabolite [20]. Fluorescent probing of rat liver native DNA using acridine orange showed that the cortisol-apoA-I complex exerted practically no effect on the DNA secondary structure, while the tetrahydrocortisol-apoA-I complex increased the number of single-stranded regions [19].

The present study reports the results of the DNA enzymatic analysis after a treatment with cortisol, tetrahydrocortisol, and their complexes with apolipoprotein A-I in vitro. S1 nuclease was used as an enzyme splitting only the single-stranded regions of nucleic acids [21, 22]. The ability of DNA with specifically changed secondary structure to participate in the transcription process was estimated from the interaction with bacteriophage T7 DNA-dependent RNA-polymerase. We evaluated also the action of cortisol, tetrahydrocortisol and their complexes with apolipoprotein A-I on a rate of protein and DNA biosynthesis in primary cultures of hepatocytes. The ability of eukaryotic DNA and synthetic duplex of $CC(GCC)_5 \cdot GG(CGG)_5$ type to interact actively with cortisol-apoA-I and THC-apoA-I complexes was studied using the method of small-angle X-ray scattering (SAXS). IR spectroscopy was used to examine the molecular mechanisms of this interaction.

# Materials and Methods

## Production of Apolipoprotein A-I and its Complexes with Hormones

The study was performed with Wistar male rats weighing 180-200 g. The animals were killed under light Nembutal anesthesia. Lipoproteins of blood serum were isolated by ultracentrifugation after chylomicrones removal from the blood serum [23]. Four main lipoprotein (LP) fractions were obtained: VLDL ($0.94 < d < 1.006$ g/ml), LDL ($1.006 < d < 1.63$ g/ml), $HDL_2$ ($1.063 < d < 1.125$ g/ml), and $HDL_3$ ($1.125 < d < 1.21$ g/ml). Further the $HDL_3$ fraction was used. Delipidation of LP was conducted with a cooled chloroform-methanol mixture (1:1), which was followed by multiple ether washing. An apo-$HDL_3$ mixture was deposited over a column (1.6×100 cm) with 6B-CL Sepharose (Pharmacia, Sweden) and eluted with 0.01 M tris-HCl buffer, pH 8.6, containing 6 M urea, 0.01% sodium azide, and 1 mM phenylmethylsulfonyl fluoride. Electroelution profile was recorded with a

UV detector 2151 LKB (Sweden). The purity of the obtained apolipoprotein A-I was tested by electrophoresis in polyacrylamide gel (PAAG) with Ds-Na. Protein bands were visualized with 0.1% Coomassie G-250 solution in the mixture of methanol and 10% acetic acid (1:1). Protein concentration in the samples was measured according to Lowry [24].

To obtain complexes of cortisol and tetrahydrocortisol with apolipoprotein A-I, their mixture (molar ratio 2:1) was kept in 0.05 M K-phosphate buffer, pH 7.4, for 5 min at room temperature. Cortisol was purchased from Merck (Germany). Tetrahydrocortisol was kindly granted by Yu.A. Pankov (Institute of Experimental Endocrinology, Endocrinologic Scientific Center, RAMS, Moscow).

## Isolation of Hepatocytes

Liver cells were isolated by the Seglen method [25] in our modification [26] using recirculating perfusion of the organ with 0.03% collagenase solution (Boehringer Mannheim, Germany). After tissue dissociation, hepatocytes were separated from nonparenchymal cells via differential centrifugation. Only hepatocytes were used in the work. The cells were counted in a Goryaev chamber. Cell viability was estimated by Trypan Blue exclusion [25, 26]. Structural integrity of the cells was determined also by light and electron microscopy. The obtained cells were resuspended in DMEM, pH 7.4, containing 15 mM HEPES, 10% fetal serum, and 50 μg/ml gentamycin. The cells were incubated in a $CO_2$ incubator (Cole Parmer, USA) for 24 h. For this purpose, 24-well plates (Linbro, USA) coated with the first type collagen were used. The final density of hepatocytes in a primary culture was 700 cells per $mm^2$.

## DNA Biosynthesis Rate Measurement [27]

The rate of DNA biosynthesis in hepatocyte culture was determined by the addition of $^3$H-thymidine at 37 kBk/ml concentration 18 hours before the end of incubation. The cell monolayer was washed from nonbound marker with a Krebs-Ringer solution, solubilized with 0.3 M NaOH, transferred to vials, where protein was precipitated using 10% trichloroacetic acid (TCA). The resulting precipitate was dissolved in 0.5 M NaOH. The 200 mcl aliquots were transferred to the membrane GF/C filters (Whatman, UK) to measure the radioactivity associated with DNA.

## Protein Biosynthesis Rate Measurement [28]

Two hours before the end of incubation, $^{14}$C-leucine (37 kBk/ml) was added to the medium. The reaction was stopped with 0.2 M NaOH. The 100 mcl aliquots from each well were transferred to the FN-16 filters (Whatman, UK) pretreated with a 0.1 M leucine solution in 10% TCA. The filters were sequentially washed with the TCA solution and ethanol-ether mixture (1:1).

Radioactivity was measured with a fluid-scintillation counter Mark-III (USA) using a toluene scintillator and expressed as impulses per minute per a well.

### Production of DNA, Enzymes and Oligonucleotides

DNA from adult rat liver was isolated according to [29] and tested by electrophoresis in 0.3% agaric gel. DNA was $26\times10^3$ kDa, about 42 kbp. DNA of phage λ and its restricted fragments Hind III were purchased from Sibenzyme (Novosibirsk, Russia), and S1 nuclease Aspergillus oryzae with specific activity 7 u./μl – from Sigma (USA). RNA-polymerase of T7 bacteriophage (specific activity 200000 u./$OE_{280}$, electrophoretic purity of the preparation 95%) was provided by V.N. Ankilova (Novosibirsk Institute of Bioorganic Chemistry, Siberian Division, Russian Academy of Sciences), cortisol – by Serva (Germany), tetrahydrocortisol was kindly granted by Yu.A. Pankov (Institute of Experimental Endocrinology, Endocrinologic Scientific Center, RAMS, Moscow).

We have shown previously that the supposed site of DNA binding to THC-apoA-I complex is the sequence of $(GCC)_n$ type [30]. Computer search over the international databases (GenBank, DNA library) showed that this sequence is frequently found in eukaryotic genes, in particular, in human genes. The number of such replicas varies over a wide range. For instance, there are 14 replicas of this type in 5′-flanking region of the gene responsible for synthesis of the human multidrug-resistance protein [31]. In the present work we used two synthetic mutually complementary sequences with replicas of $CC(GCC)_5$ and $GG(CGG)_5$ types, and their duplex. For comparison purposes, a fragment of kDNA-regulatory region of cytochrome P-450 gene and polyT-oligonucleotide were chosen. Oligonucleotides were synthesized at the State Research Center of Virology and Biotechnology VECTOR, RF Ministry of Health. The primary structures of single-stranded oligonucleotides were as follows:

1. 5′-CCGCCGCCGCCGCCGCC-3′;
2. 5′-GGCGGCGGCGGCGGCGG-3′;
3. 5′-ATCTTTAACTGATGAACTTCT-3′;
4. 5′-TTTTTTTTTTTTTTTTTTT-3′.

To prepare a duplex from monomers, solutions of oligonucleotides 1 and 2 were mixed in equimolar ratio 1:1, the mixture was heated to 70°C and cooled slowly (annealing) for 2 h to room temperature [32].

## S1 Nuclease Analysis

To obtain apoA-I-hormone (cortisol, tetrahydrocortisol-THC) complex, 1.38 ng of hormone was added to 57 ng of apoA-I in 10 mM Tris-HCl buffer, pH 8.6, mixed and kept at room temperature for 5 min [19]. The protein/hormone molar ratio was 1:2, and in the other experiments the ratio was 3:1, 1:1 or 1:3. Then the obtained hormone-apoA-I complex was added to 250 ng of the native DNA dissolved in water and kept at room temperature for 10 min. The 5-fold buffer for S1 nuclease was added to a mixture up to the concentrations [21]: 33 mM NaAc, 50 mM NaCl, 1 mM $ZnSO_4$, pH 5, in the ratio of 5 protein molecules (in the composition of apoA-I-THC complex) per 1 kbp (in the other experiments the ratio was 1 protein molecule per 1 kbp or 50 protein molecules per 1 kbp), mixed, and kept for 5 min; 0.2 u. S1 nuclease was added to the resulting solution and incubated for 30 min at 37°C. Composition of the reaction mixtures was as follows: 15 μL volume, 250 ng DNA, apoA-I from 2.27 to 5.70 ng, hormone (THC, cortisol) from 0.028 to 20.7 ng, S1 nuclease 0.2 u., 3 μL of the 5-fold buffer for S1 nuclease, and water. The obtained DNA preparations were tested by electrophoresis in 0.5 and 1% agaric gel [22]. DNA of phage λ and its restricts Hind III were used as tags for estimation of DNA fragments sizes. After electrophoresis, the gel was dyed in a solution of 0.5 μg/ml ethydium bromide, and its UV plots were taken.

To evaluate quantitatively the effect of hormones and their complexes with apoA-I, the sections of electrophoregrams with the split DNA fragments of size less than 4.3 kbp were taken. Scanning of these photofilm sections showed a direct proportion between the number of λ DNA tag restricts Hind III and the area of optical density profile of the corresponding section of negative. The BandScan program was used to estimate the amount of split nucleotide material with respect to the standard sample of specified weight.

## Small-angle X-ray Scattering

The SAXS method was used to estimate the interaction of eukaryotic DNA and synthetic oligonucleotides with cortisol-apoA-I and THC-apoA-I complexes [33]. Small-angle X-ray patterns were measured with a Siemens diffractometer (Germany). X-ray wavelength $\lambda$ = 0.154 nm ($CuK_\alpha$). Homogeneous preparations of apoA-I, DNA and oligonucleotides with the initial concentrations of 0.40, 3.12 and 1.52 mg/ml, respectively, were used. Upon taking the SAXS plots, the temperature of the initial samples was 20°C. To study the interaction of oligonucleotides with apoA-I, the mixtures of oligonucleotide solutions with apoA-I and its cortisol and tetrahydrocortisol complexes of various concentrations were prepared. Small-angle X-ray patterns were taken over the angle range of $0.0245 \le h \le 3.423\ nm^{-1}$, where $h = 4\pi \cdot Sin(\theta)/\lambda$, and $2\theta$ is the scattering angle. The SAXS plots were corrected for background scattering and collimation and smoothed. Reliability of the obtained data was estimated using the Student's t-criterion at the significance level of $P < 0.05$.

## IR Spectroscopy

The mechanism of interaction of glucocorticoids (tetrahydrocortisol, cortisol and their apoA-I complexes) with oligonucleotide $CC(GCC)_5 \cdot GG(CGG)_5$ (duplex) was studied using infrared (IR) spectroscopy. Solutions of hormones were prepared with 0.01 M tris-HCl buffer, pH 7.2, with the addition of $EDTANa_2$. The concentration of oligonucleotide was $4 \cdot 10^{-4}$ M. Tetrahydrocortisol was dissolved in the Merck methanol to provide the concentration of $10^{-2}$ M, then it was diluted by 4 orders of magnitude ($C_{THC} = 2.0 \cdot 10^{-6}$ M) with 0.001 M tris-HCl buffer, pH 7.2.

A film for the IR spectra measurement was prepared in a cuvette with fluorite backing via slow evaporation of water under weak vacuum at a pressure of ca. 0.1 atm (ca. $0.5 \cdot 10^4$ n/m$^2$) and temperature of 4±1°C [34]. Drying was conducted for 150 min. The volume of oligonucleotide solution poured into a cuvette was 20 μl. 20 μl of $H_2O$, the same amount of tris-buffer, $EDTANa_2$, and 5 μl of hormone were added to the solution. Stirring and incubation lasted 10 min at room temperature. A cuvette was placed horizontally on a special stage of a vacuum unit.

The concentration of apoA-I in the experiment with THC or cortisol and oligonucleotide comprised $2 \cdot 10^{-6}$ M. Dialysis of apoA-I was performed against 0.001 M Tris-HCl buffer, pH 7.2, with a sixfold buffer replacement at T = 4°C.

After the film preparation, the cuvette was placed into an optical chamber and blown with dry air for 30 min. Then the scanning was turned on. IR spectra were taken with a Specord-M80 spectrometer (Leipzig, Germany) sequentially: experiment and control relative to fluorite backing, or experiment – control when taking the differential spectrum. Smoothing, integration, determination of band frequency, and mathematical processing were performed using special programs recorded on the Data Handling cassette furnished as part of device.

# Results

## Effect of Tetrahydrocortisol, Cortisol and their Complexes with Apolipoprotein A-I on Protein and DNA Synthesis in Primary Culture of Hepatocytes

Blood lipoproteins are of great interest as a transport form of steroid hormones. This relates primarily to the hormones that are capable of binding to the protein component of lipoproteins. Previously we have shown that apolipoprotein A-I forms a rather firm complex with corticosteroids. For instance, $K_{ass}$ of apoA-I and corticosterone comprises $(0.4 \pm 0.1) \cdot 10^6$ $M^{-1}$. In HDL the value of $K_{ass}$ for corticosterone is even higher, attaining $(2.0 \pm 0.2) \cdot 10^6$ $M^{-1}$ [13]. Note, for comparison, that $K_{ass}$ for transcortin is $3 \cdot 10^7$ $M^{-1}$ [35]. The transport role of apoA--I for steroid hormones is of particular interest, since this protein easily enters not only a cell but also a nucleus through the corresponding receptor [14]. In the present work we tried to estimate the biochemical and physiological role of a new molecular complex. For this purpose, we used cortisol (an active glucocorticoid) and tetrahydrocortisol, which was previously considered to be an inactive hormone species. The rate of DNA and protein biosynthesis in primary culture of rat hepatocytes was determined.

Table 1 presents the results of these experiments. Cortisol was shown to exert no effect of the rate of $^{14}$C-leucine incorporation in protein, while tetrahydrocortisol and apoA-I increased this rate only slightly. Cortisol-apoA-I complex slightly decreased the rate of protein biosynthesis, while THC-apoA-I complex was shown to increase it sharply. Similar data were obtained for the rate of DNA biosynthesis. It was found that cortisol, THC and apoA-I virtually did not affect the DNA biosynthesis. Cortisol-apoA-I complex slightly inhibited the process, while THC-apoA-I complex considerably enhanced DNA biosynthesis in hepatocytes.

**Table 1. Effect of cortisol, tetrahydrocortisol and their complexes with apoA-I on protein and DNA synthesis in primary culture of hepatocytes**

| Experimental conditions | Rate of $^{14}$C-leucine incorporation in protein, pulse/min per a well | Rate of $^{3}$H-thymidine incorporation in DNA, pulse/min per a well |
|---|---|---|
| Control | 3492 ± 171 | 7653 ± 127 |
| Cortisol | 3190 ± 62 | 7406 ± 110 |
| Tetrahydrocortisol | 3630 ± 122# | 7305 ± 150 |
| Apolipoprotein A-I | 3840 ± 93$^{*\#}$ | 7200 ± 121 |
| Cortisol-apoA-I | 3083 ± 60$^{*}$ | 7336 ± 194 |
| THC-apoA-I | 4860 ± 503$^{*}$ | 8117 ± 140* |

* — $p < 0.05$ with respect to control,
# — $p < 0.05$ with respect to hormone-apoA-I complex

Therefore, it can be concluded that apoA-I acts as a transport form of hormones in primary culture of hepatocytes. Note that being a part of complexes with apoA-I, cortisol did not activate DNA and protein biosynthesis, while THC greatly increased the rate of the both processes. This essential difference in the molecular mechanism of cortisol and THC action can be attributed only to the presence of reduced $\Delta^4$,3-ketogroup in THC molecule. Previously we have pointed out this important fact, since OH-group in the third position of the hormone A ring may competitively participate in the formation of hydrogen bonds at the interaction of hormone-apoA-I complex with GC pairs in DNA structure, leading to its local melting [33]. In these studies, apolipoprotein A-I is shown to be a specific carrier of steroid hormones, which is associated with a fundamentally different mechanism of steroid hormones involvement in the regulation of gene expression as compared to the known mechanisms. The analysis of this mechanism is presented below.

## Effect of Tetrahydrocortisol, Cortisol and their Complexes with Apolipoprotein A-I on the Secondary Structure of Eukaryotic DNA (Enzymatic Analysis)

The native DNA of rats is known to have some number of single-stranded regions. In mature rats, the number of such regions is doubled as compared to that in newborn rats [36]. In the present study, DNA was isolated from mature rat liver with retaining nativity. Actually, the treatment of initial DNA with S1 nuclease reveals some number of single-stranded fragments (Figure 1, track 2).

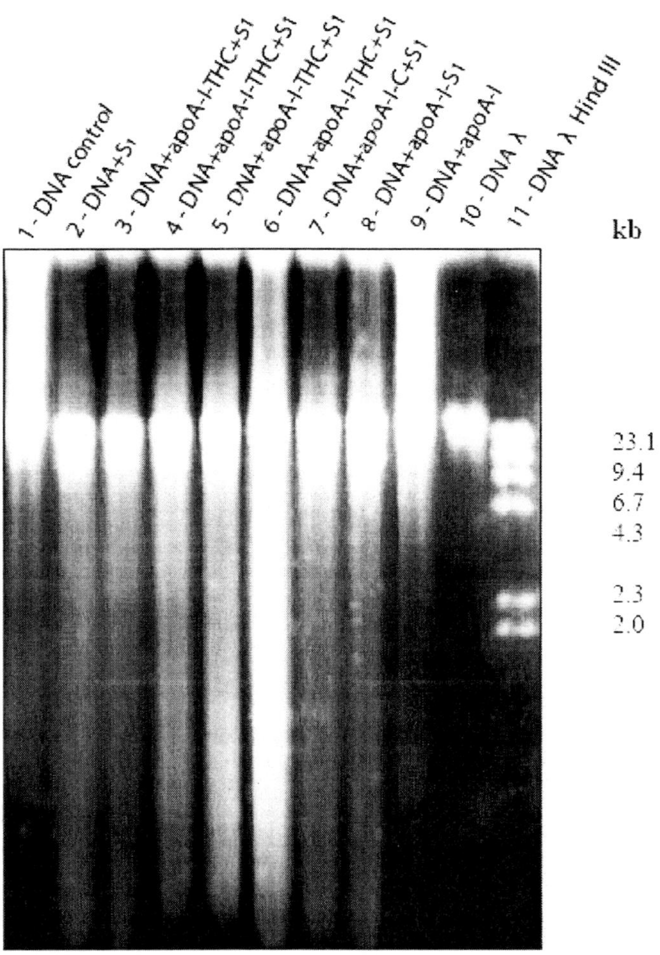


Figure 1. DNA separation by 1% agaric gel electrophoresis. *1* – DNA control, *2-8* – DNA treated with S1 nuclease (8u/mL): *3-6* – DNA+[apoA-I-THC], 1 protein molecule per 5 kb, 1 protein molecule per 1 kb, 5 protein molecules per 1 kb, and 10 protein molecules per 1 kb; *7* - DNA+[apoA-I-cortisol], 10 protein molecules per 1 kb; *8* - DNA+apoA-I, 5 protein molecules per 1 kb; *9* - DNA+apoA-I (without S1 nuclease), 5 protein molecules per 1 kb; *10* – DNA λ Hind III; *11* - DNA λ.

A considerable increase in the number of S1-nuclease-sensitive regions in the native DNA isolated from rat liver was observed after the action of apoA-I-tetrahydrocortisol complex (Figure 1, tracks 3-6). Complex of apoA-I with cortisol did not produce such effect, i.e., the number of S1-nuclease-sensitive fragments remained comparable with that in the initial DNA (Figure 1, compare tracks 6, 7 and 2). The treatment of DNA merely with apoA-I also did not cause a significant change in the number of single-stranded fragments.

A broad spectrum of sizes of DNA fragments split by S1 nuclease (from 15 to 0.5 kbp) indicates the irregular distribution of the interaction sites of rat DNA and THC-apoA-I complex. The exception is the highly specific sites, whose splitting leads to fragments of size 5.5 kbp. Scanning the electrophoregram tracks from photofilms allows estimating the number of DNA fragments formed under the action of S1 nuclease before and after the introduction of hormone-apoA-I complex (Figure 2). The total weight of DNA applied over a DNA track was 500 ng. With the complex/DNA ratio equal to 1, 2 and 5 per 1 kbp, the number of split DNA fragments increased reliably (Figure 2, columns 3, 4 and 5, $p < 0.05$). An additional increment in the number of split DNA fragments, without regard to their initial content, comprised respectively ~5% of the total DNA at the ratio of 1 protein molecule per 1 kbp, ~12% at 2 protein molecules per 1 kbp, and ~25% at 5 protein molecules per 1 kbp. Therefore, in the case of specific action of the complex, no less than 5% of native DNA is split, which corresponds to 2 split sites of complex binding per 100 kbp. Further increase in the amount of split DNA under the action of high concentrations of the complex may be caused by its unspecific binding to the single-stranded regions that form near the primary specific sites of binding, which may lead to a pronounced untwisting of the duplex. However, at the ratio of 10 and more apoA-I protein molecules (in THC-apoA-I complex) per kbp, inhibition of the S1 nuclease action on DNA is observed. This effect is due to the fact that the excess of complex, upon its binding to single-stranded regions, shields them from the action of S1 nuclease.

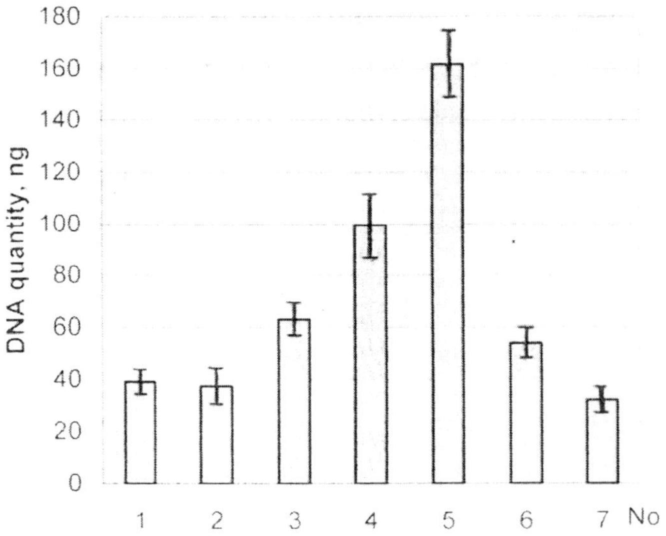


Figure 2. Quantity of S1-cleavage DNA < 4,3 kb under the apoA-I THC complex concentrations (n=5, P<0,05); S1 nuclease 8u/mL. *1* – DNA control,; *2-5* - DNA+[apoA-I-THC] (0.2, 1, 2, and 5 protein molecules per 1 kb, respectively); *6* - DNA+[apoA-I-cortisol (5 protein molecules per 1 kb); *7* - DNA+apoA-I (5 protein molecules per 1 kb).

Along with irregular distribution of the sites of DNA interaction with the complex, we found the replicas of size 6 and 5.5 kbp (Figure 3). To estimate their increment under the action of THC-apoA-I complex, the calculations similar to those reported above were performed. For standardization of DNA fragment weights, the fragment of λ DNA Hind III of size 6.7 kbp was chosen. In these experiments, the amount of complex varied from 1 apoA-I protein molecule per 5 kbp of DNA to 50 protein molecules per 1 kbp. A 6 kbp fragment appeared under the action of S1 nuclease in all cases. No significant difference in the amount of this fragment in DNA was observed in the experiments at different complex/DNA ratios. On the average, its amount comprised 5.5±0.9 ng (with 24 measurements).

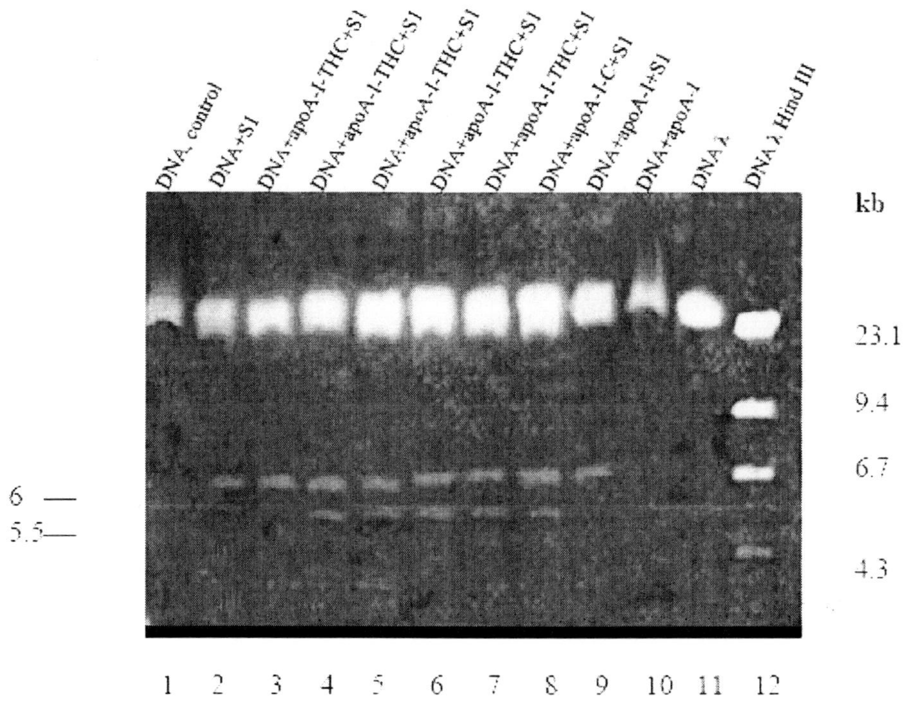


Figure 3. DNA separation by 0,5% agaric gel electrophoresis. DNA was treated with complex apoA-I-THC. *1* – DNA control, *2-9* – DNA treated with S1 nuclease (8u/ml): *3-7* – DNA+[apoA-I-THC], 1 protein molecule per 5 kb, 1 protein molecule per 1 kb, 5 protein molecules per 1 kb, 10 protein molecules per 1 kb, and 50 protein molecules per 1 kb; *8* - DNA+[apoA-I-cortisol], 5 protein molecules per 1 kb; *9* - DNA+apoA-I, 5 protein molecules per 1 kb; *10* - DNA+apoA-I (without S1 nuclease), 1 protein molecule per 1 kb; *11* - DNA λ.; *12* – DNA λ Hind III.

After preincubation of DNA with hormone-apoA-I complex (THC, cortisol), a 5.5 kbp fragment appeared under the action of S1 nuclease. Although the complex/DNA ratio varied greatly, the absence of this fragment was reliably ascertained only for the reaction mixture with the ratio of 1 protein molecule (in the composition of THC-apoA-I complex) per 5 kbp. On the average, the DNA amount in this fragment was 2.6±0.6 ng (with 15 measurements). From the equation $N \times d/D = m/M$, where N is the number of the fragment replicas in the genome, d is the size of the fragment, bp, D is the size of the genome, bp, m is the weight of the fragment DNA, and M is the weight of total DNA in the reaction mixture, one may estimate the number of replicas of these fragments. It was taken into account that the size of rat genome is $3 \times 10^9$ bp. The number of replicas of the 6 kbp fragment in the rat genome was found to be ~5000, while that of the 5.5 kbp fragment ~2500.

Of interest is the minimum size spread of fragments composing the both replicas. Many replicating genes are known to have S1-nuclease-sensitive regions that occur also in the promoter area. For the most frequent genes, for instance, ribosomal RNA, the number of replicas does not exceed some hundreds per genome. For the fragments of size 6 and 5.5 kbp found in our experiments, the number of replicas is much greater: ~5000 and ~2500 replicas per genome, respectively. When considering a possible role of these fragments, it should be taken into account that apoA-I was detected in hepatocytes nuclei, and in large amount in the nuclear matrix [15]. At the same time, several proteins of the nuclear matrix are known to be capable of specific binding to satellite DNA [37]. Some structural properties of satellite DNA imply hypersensitivity to S1 nuclease [38]. Binding of chromosome DNA regions hypersensitive to S1 nuclease to the nuclear matrix was demonstrated in the works by Svetlova et al. [39]. In eukaryotic genome, the total number of sites of DNA binding to the nuclear matrix does not exceed 10 000, which corresponds approximately to 1% of the genome [40]. Therefore, if the fragments of size 5.5 and 6 kbp belong to the satellite DNA, it can be supposed that some fraction of S1-nuclease-sensitive regions forms under the action of the complex under consideration at the sites of satellite DNA binding to the nuclear matrix. Certainly, this assumption needs further experimental verification.

## Interaction of Tetrahydrocortisol-apolipoprotein A-I and Cortisol-Apolipoprotein A-I Complexes with Eukaryotic DNA (SAXS Method)

Molecular mechanism of DNA interaction with apoA-I and its complexes with cortisol and tetrahydrocortisol was studied in our work using the SAXS method. Concentrations of the initial components in four mixtures are listed in Table 2. Figure 4a displays the SAXS patterns obtained from the DNA preparation and one of DNA mixtures with apoA-I (mixture 4) in the presence and in the absence of hormones. The experimental SAXS data were used to calculate the values of radius of gyration ($R_g$) and intensity of X-ray zero angle scattering [I(0)] for all four mixtures of DNA with apoA-I (Table 2). One may see from Figure 4a and Table 2 that DNA interaction with apoA-I, at a given accuracy of the SAXS intensity measurement, was observed only in the presence of THC and led to the formation of composite complexes $P_nS$. When THC was replaced by cortisol or in the absence of hormones, no interaction of apoA-I with DNA was observed. Within the accuracy of measurement, the size of $P_nS$ complexes corresponds to the size of DNA molecules in a free state ($R_g$ = 39.0 ± 0.5 nm); changes are found only in the values of I(0) related to the concentration and molecular weight of the complexes [32, 41]:

$$I(0) \sim N \cdot (Mk)^2, \qquad (1)$$

where N and Mk are, respectively, the number and molecular weight of $P_nS$ complexes involved into scattering.

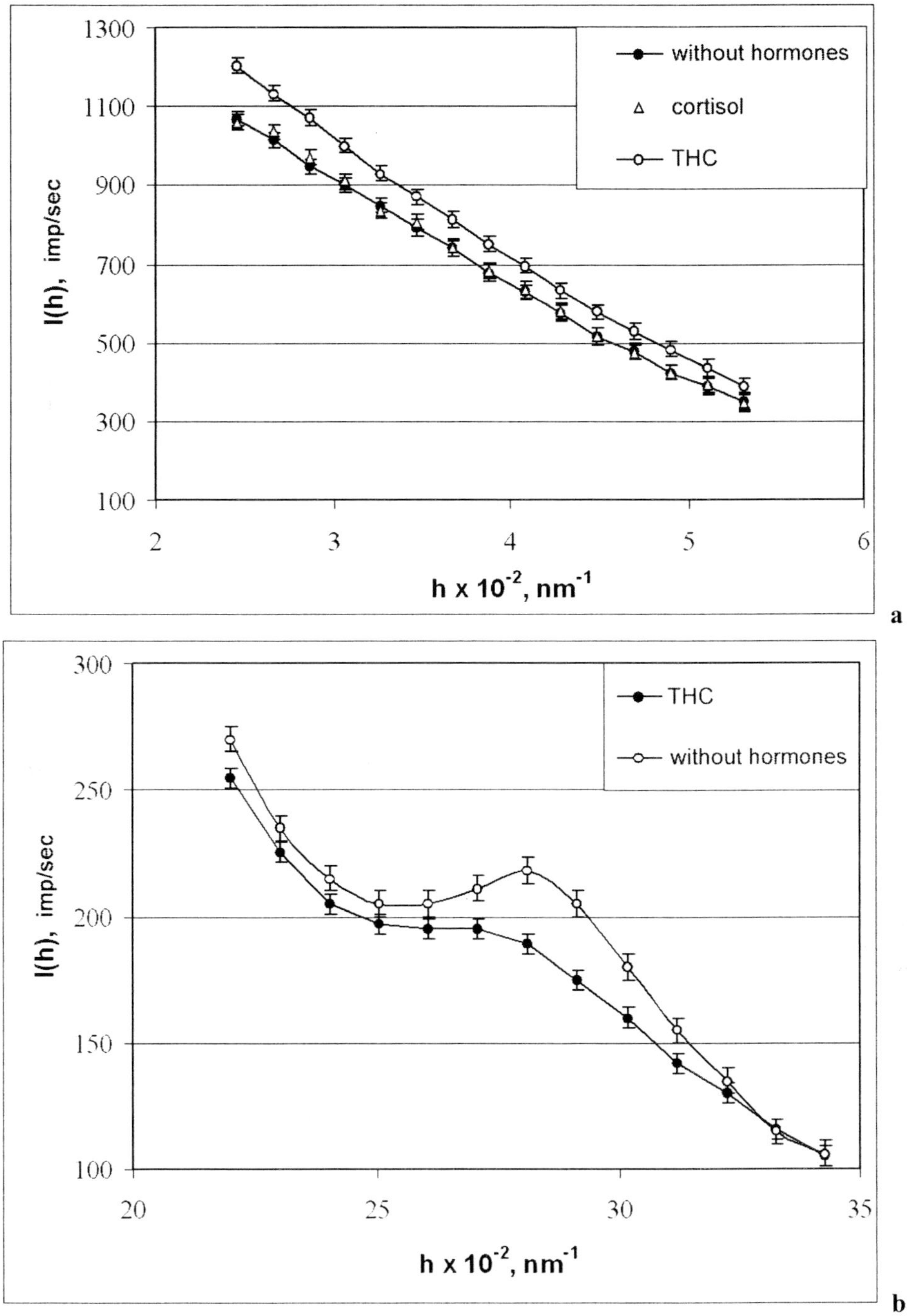


Figure 4. SAXS patterns obtained from DNA preparations and one of four mixtures with apoA-I (mixture 4): in the Guignet region **(a)**; in the region of diffraction peak at h ≈ 2.66 $nm^{-1}$ **(b)**. Concentrations of protein and DNA in mixture 4 are presented in Table 2.

Table 2 presents the calculation results for parameter n and the estimates of $C_P/C_S$ balance for all DNA mixtures with apoA-I in the presence of THC used in the study. One may see that the n value in the mixtures does not exceed the values of the balance and in mixtures 3-4 attains its maximum, which corresponds to a 54:1 stoichiometry of the forming $P_nS$ complex upon DNA interaction with THC-apoA-I complex. Of course, this estimate of stoichiometry is very high and reflects not only specific but also nonspecific interactions of completely deproteinized DNA with THC-apoA-I complex.

**Table 2. Mean values of radius of gyration ($R_g$), intensity of X-ray zero angle scattering I(0), and the number of apolipoprotein A-I molecules (n) in the $P_nS$ complexes obtained from the SAXS data [16] at molar concentrations of protein ($C_P$) and DNA ($C_S$) in mixtures 1-4**

| Mixture | $C_P$, μM | $C_S$, μM | $R_g$, nm | I(0), cps | N | $C_P/C_S$ |
|---|---|---|---|---|---|---|
| 1 | 1.07 | 0.04 | 38.6 ± 0.5 | 1516 ± 9 | 26 | 27 |
| 2 | 2.14 | 0.04 | 39.1 ± 0.6 | 1580 ± 8 | 47 | 54 |
| 3 | 3.21 | 0.04 | 38.9 ± 0.5 | 1601 ± 10 | 55 | 81 |
| 4 | 5.35 | 0.04 | 38.8 ± 0.6 | 1596 ± 11 | 53 | 133 |

Note: For DNA (at CS = 0.04 μM), I(0) = 1430 ± 10 cps.

Where does the interaction of THC-apoA-I complex with DNA lead? First of all, there are some changes in DNA secondary structure. It is known that the integrity of DNA secondary structure can be judged from the occurrence of diffraction maximum in the angle range of h ≈ 2.66 $nm^{-1}$ [32, 42] in the SAXS plot corresponding to the diameter of DNA double chain (2.4 nm). Figure 4b shows that at h ≈ 2.7 $nm^{-1}$ the diffraction maximum is clearly observed in the SAXS plots of native DNA preparations and DNA mixtures with apoA-I without THC, while with similar mixtures in the presence of THC the peak value decreases reliably, indicating the rupture of native DNA secondary structure upon DNA interaction with THC-apoA-I complexes. Such interaction resembles the enzyme–substrate interaction, where THC acts as a cofactor, leading to the rupture of hydrogen bonds between the pairs of nitrous bases.

Further we applied the SAXS method in the experiments with single-stranded oligonucleotides, whose structure is presented in the Materials and Methods Section. Figure 5 shows the SAXS plots taken from the solutions of single-stranded mutually complementary oligonucleotides 1 and 2 used in the study. The radius of gyration for these oligonucleotides were calculated from the SAXS data: $R_g$ = 1.24 ± 0.05 nm (oligonucleotide 1) and $R_g$ = 1.21 ± 0.06 nm (oligonucleotide 2). Within the accuracy of measurement, $R_g$ values for these oligonucleotides are virtually identical. Assuming that DNA mean density is 1.55 $g/cm^3$ and molecular weight M ≈ 5700 Da, one may estimate mean volume of oligonucleotides 1 and 2: V = 6.31 $nm^3$.

To estimate by SAXS the interaction of apoA-I with oligonucleotides, the equilibrium mixtures of oligonucleotides and apoA-I complexes with THC and cortisol were used at the first stage. The mixtures were prepared: four mixtures of apoA-I with oligonucleotide 1 and two mixtures for estimating the effects of apoA-I interaction with oligonucleotides 3 and 4 and the effect of THC and cortisol. Mean values of the radius of gyration ($R_g$) and differential intensities of the zero angle scattering ($\Delta I(0)$) were calculated from the SAXS data for all apoA-I mixtures with oligonucleotides and glucocorticoids (see Table 3).

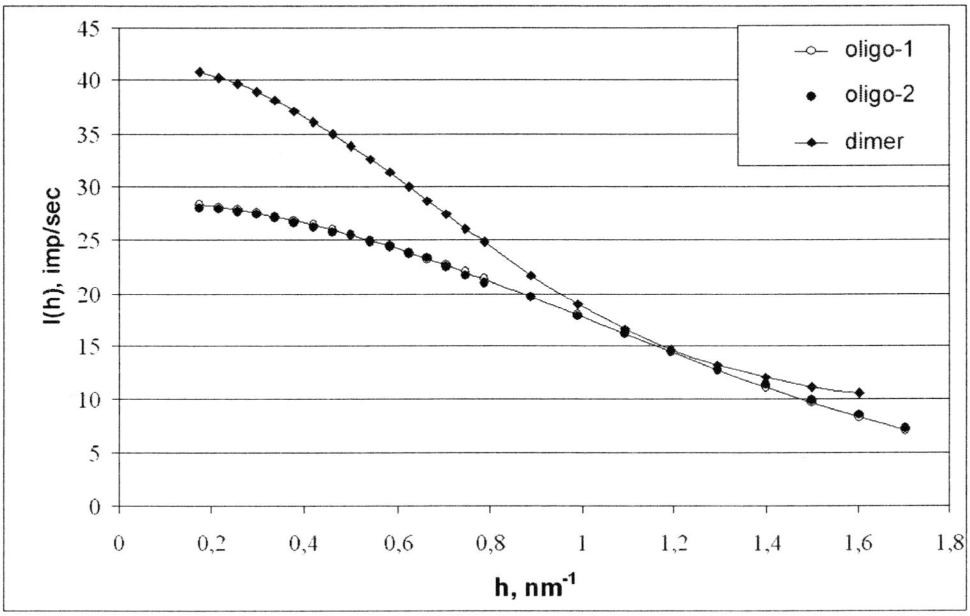


Figure 5. SAXS patterns obtained from solutions of oligonucleotides 1 and 2 (concentration 1.5 mg/ml) and duplex (concentration 2.2 mg/ml). Algorithms presented in [32] were used to introduce corrections for background scattering and collimation of X-ray beam.

**Table 3. Mean values of radius of gyration ($R_g$), experimental ($\Delta I(0)$) and model ($\Delta J(0)$) differential intensities of X-ray scattering to the zero angle obtained from the SAXS data at molar concentrations of protein ($C_P$) and oligonucleotide ($C_S$) in the mixtures**

| Mixture | Oligonucleotide | $[P_o]$, μM | $[S_o]$, μM | $R_g$, nm | ΔI(0), cps |
|---|---|---|---|---|---|
| 1 | Oligo-1 | 6.6 | 51.5 | 2.25 ± 0.04 | 3.2 ± 0.3 |
| 2* | Oligo-1 | 6.6 | 51.5 | 2.33 ± 0.05 | 1.5 ± 0.3 |
| 3** | Oligo-1 | 6.6 | 51.5 | 2.28 ± 0.04 | 1.1 ± 0.2 |
| 4 | Oligo-1 | 9.9 | 25.7 | 2.32 ± 0.05 | 2.9 ± 0.4 |
| 5 | Oligo-3 | 9.9 | 25.7 | 2.35 ± 0.05 | 0.4 ± 0.3 |
| 6 | Oligo-4 | 9.9 | 25.7 | 2.37 ± 0.05 | 0.3 ± 0.2 |

Note: In mixtures 1, 4-6 THC is present with concentration $2{\cdot}10^{-4}$ M.

* In mixture 2 hormones are absent.

** In mixture 3 cortisol is present with concentration $2{\cdot}10^{-4}$ M.

Primary structures of single-chain oligonucleotides: oligo-1 (oligonucleotide 1), oligo-3 (oligonucleotide 3), and oligo-4 (oligonucleotide 4) are presented in the Materials and Methods Section.

One may see from Table 3 that, with a given accuracy of the SAXS intensity, a noticeable interaction of apoA-I was observed only with oligonucleotide 1. In the absence of THC, the interaction of oligonucleotide with apoA-I was considerably weaker. More detailed titration data were obtained to study the mechanism of THC-apoA-I interaction with oligonucleotides 1 and 2 (Table 4).

**Table 4. Values of differential ($\Delta I(0)$) and model ($\Delta J(0)$) intensities of X-ray scattering to the zero angle obtained from the SAXS data at molar concentrations of protein ($C_P$) and oligonucleotides 1-2 ($C_S$) in the mixtures**

| Mixture | $[P_o]$, μM | $[S_o]$, μM | Oligo-1 | | Oligo-2 | |
|---|---|---|---|---|---|---|
| | | | $\Delta I(0)$, cps | $\Delta J(0)$, cps | $\Delta I(0)$, cps | $\Delta J(0)$, cps |
| 1 | 6.1 | 44.2 | 2.5 ± 0.2 | 2.45 | 2.7 ± 0.2 | 2.55 |
| 2 | 9.2 | 22.1 | 2.2 ± 0.1 | 2.21 | 2.3 ± 0.2 | 2.30 |
| 3 | 10.5 | 12.3 | 1.5 ± 0.1 | 1.52 | 1.6 ± 0.1 | 1.62 |
| 4 | 11.2 | 7.4 | 1.1 ± 0.1 | 1.02 | 1.1 ± 0.1 | 1.07 |

Note: Primary structures of single-chain oligonucleotides 1 and 2 are presented in the Materials and Methods Section.

The changes in experimental values of $\Delta I(0)$ shown in Table 4 indicate the formation of complexes of initial macromolecules to occur in the mixtures [42, 43, 44]. The values of $\Delta I(0)$ are related to concentration and molecular weight of $P_mS_n$ complexes similar to (1). Here P denotes the complexes of apoA-I with THC; S is the oligonucleotide; m, Mp, n, and Ms are the amounts and molecular weights of protein molecules (with THC) and oligonucleotides aggregated into $P_mS_n$ complexes. To estimate the maximum stoichiometry of complexes formed due to the interaction of oligonucleotides (S) with apoA-I (P), we used a highly cooperative equilibrium scheme and the methodology developed earlier [42, 43, 44]:

$$\begin{aligned}
&\overset{K_{mn}}{m \cdot P + n \cdot S \Leftrightarrow PmSn}, \qquad (4)\\
&K_{mn} = [P]^m \cdot [S]^n / [PmSn],\\
&R(m, n, K_{mn}) = 2 / p \cdot \Sigma\, |\Delta I_i(0) - \Delta J_i(0)| / (\Delta I_i(0) + \Delta J_i(0)),\\
&\Delta I_i(0) = I_i(0) - [P_o] \cdot J_P(0) - [S_o] \cdot J_S(0),\\
&\Delta J_i(0) = [P_mS_n] \cdot (J_{mn}(0) - m \cdot J_P(0) - n \cdot J_S(0)),\\
&J_{mn}(0) = (m \cdot (J_P(0))^{1/2} + n \cdot (J_S(0))^{1/2})^2,
\end{aligned}$$

where $K_{mn}$ is the dissociation constant for $P_mS_n$ complex; $[P_o]$, $[S_o]$, $[P]$, $[S]$, $[P_mS_n]$, $J_P(0)$, $J_S(0)$, $J_{mn}(0)$ are the initial and equilibrium molar concentrations and intensities (at h = 0) of protein, apoA-I, oligonucleotides and their complexes in titration mixtures, respectively; R is the optimization criterion; $I_i(0)$ is the intensity of scattering of the i-th titration mixture; $\Delta I_i(0)$ and $\Delta J_i(0)$ are the experimental and simulated values of differential intensities of scattering (at h = 0) of the i-th mixture [42, 43, 44].

One may see from Table 5 that the minimum values of criterion R correspond to the 1:1 stoichiometry of the formed complexes (m = 1, n = 1) at association constant $K_{11} = 65.0 \pm 3.5$ μM ($K_{ass11} = 1.54 \cdot 10^4$ $M^{-1}$) for oligo-1, and $K_{11} = 69.8 \pm 4.5$ μM ($K_{ass11} = 1.43 \cdot 10^4$ $M^{-1}$) for oligo-2. Therefore, single-stranded oligonucleotides 1 and 2 form in the mixtures with apoA-I in the presence of THC complexes with stoichiometry 1:1 at nearly identical association-dissociation constants. In addition to experimental values, Table 4 lists the theoretical (model) values of the interaction characteristics obtained from the apoA-I preparation and its mixtures with oligonucleotides in THC presence. Experimental SAXS data are described by model (4) within the 2.8% error for oligo-1 and 2.1% for oligo-2.

**Table 5. Values of optimization criterion R (%) for matrices with m,n = 1,3 calculated from the SAXS data obtained for the mixtures of oligonucleotides 1 and 2 with THC-apoA-I complex**

| m / n | **Oligonucleotide 1** | | |
|---|---|---|---|
| | 1 | 2 | 3 |
| 1 | 2.8 ($K_{11}$ = 65.0) | 47.2 | 76.9 |
| 2 | 12.0 | 37.4 | 78.8 |
| 3 | 26.7 | 46.9 | 57.6 |
| | **Oligonucleotide 2** | | |
| 1 | 2.1 ($K_{11}$ = 69.8) | 45.1 | 74.8 |
| 2 | 13.5 | 35.4 | 75.2 |
| 3 | 28.2 | 42.6 | 56.1 |

$J_P(0)$ = 1.68 pulse · $s^{-1}$ · $\mu M^{-1}$;
$J_S(0)$ = 0.171 pulse · $s^{-1}$ · $\mu M^{-1}$;
$J_{11}(0)$ = 2.923 pulse · $s^{-1}$ · $\mu M^{-1}$.

To estimate the interaction of THC-apoA-I complex with DNA double-stranded region (duplex) containing specific sequence of $CC(GCC)_5 \cdot GG(CGG)_5$ type, a solution of dimer molecules was prepared from monomers (see the Materials and Methods Section). Figure 5 displays the SAXS plot of the so formed duplex comprising 17 nucleotide pairs. The $R_g$ value determined from the SAXS data for dimer was higher as compared to monomer and attained $R_g$ = 1.71 ± 0.07 nm. Assuming that mean DNA density is 1.55 g/cm$^3$ and molecular weight M ≈ 11400 Da, we estimated mean volume of dimer molecules: V = 12.62 nm$^3$.

It is known that the structure of short oligonucleotides in solution is of the ellipsoid of revolution or cylinder type [41]. It can be demonstrated that the values $R_g$ = 1.2 ± 0.06 nm and V = 6.31 nm$^3$ obtained in our study for oligonucleotides 1 and 2 correspond to the structure of a cylinder with 0.727 nm radius and 3.8 nm height, while the values $R_g$ = 1.71 ± 0.07 nm and V = 6.62 nm$^3$ obtained for the duplex correspond to the structure of a cylinder with 0.85 nm radius and 5.54 nm height. Hence it follows that the duplex structure corresponds closely to the DNA region composed of 17 nucleotide pairs (length of the region is 3.4·1.7 = 5.8 nm [32]). The difference in the length of equivalent cylinder for oligonucleotides 1 and 2 can be attributed to a lower rigidity of single-stranded oligonucleotides as compared to duplexes and a greater dissimilarity of their structure to uniform cylinders. These conclusions agree well with theoretical and experimental data concerning the canonical structure of DNA [32].

The effects of interaction of apoA-I with the obtained duplex were estimated also using the equilibrium mixtures of this double-stranded oligonucleotide (SS) and apoA-I complexes with THC and cortisol, and the methodology (4) developed previously [42, 43, 44]. We prepared four new mixtures via titration of apoA-I with SS duplex in the presence of THC, and four similar mixtures in the presence of cortisol. Table 6 presents the mean values of differential intensities of zero angle scattering ($\Delta I(0)$) obtained from the SAXS data for all eight apoA-I mixtures with SS duplex and glucocorticoids.

**Table 6. Values of differential intensities of X-ray scattering to the zero angle (h = 0) obtained from experimental ($\Delta I(0)$) and calculated (model) ($\Delta J(0)$) SAXS data at molar concentrations of protein ($C_P$), monomers of oligonucleotides 1-2 ($C_S$), and duplex ($C_{SS}$) in the mixtures with THC and cortisol**

| Mixture | $[P_o]$, μM | $[SS_o]$, $[S_o]$, μM | Mixtures with THC | | | Mixtures with cortisol | | |
|---|---|---|---|---|---|---|---|---|
| | | | $\Delta I(0)$, cps | | $\Delta J(0)$, cps | $\Delta I(0)$, cps | | $\Delta J(0)$, cps |
| 1 | 5.0 | 61.1 | −13.5 ± 0.8 | | | 5.9 ± 0.4 | | 6.16 |
| | | 122.2 | | 7.4 ± 0.8 | 7.86 | | 26.8 ± 2.4 | |
| 2 | 7.5 | 30.5 | −1.1 ± 0.2 | | | 6.1 ± 0.6 | | 6.28 |
| | | 61.0 | | 9.4 ± 0.2 | 9.28 | | 16.5 ± 1.6 | |
| 3 | 8.6 | 17.1 | 2.4 ± 0.3 | | | 4.8 ± 0.4 | | 4.74 |
| | | 34.2 | | 8.3 ± 0.3 | 8.26 | | 13.3 ± 1.4 | |
| 4 | 9.2 | 10.0 | 3.4 ± 0.4 | | | 3.3 ± 0.3 | | 3.24 |
| | | 20.0 | | 6.5 ± 0.4 | 6.54 | | 6.4 ± 0.6 | |

As seen from Table 6, the values of $\Delta I(0)$ were calculated from the experimental data for two concentrations corresponding to duplex ($C_{SS}$) and single-stranded oligonucleotides 1 and 2 ($C_S$). The reason is that, unlike the mixtures with cortisol, in the presence of THC we obtained reliably negative values of $\Delta I(0)$ for two first mixtures. However, it is known that $\Delta I(0)$ cannot be negative if the interaction of macromolecules in binary mixtures leads to their complexation; and vice versa, negative values of $\Delta I(0)$ indicate that the interaction mechanism involves stages of decreasing molecular weights, i.e., implying the decomposition of initial macromolecules [42, 43, 44].

Taking into account the above data on the interaction of THC-apoA-I complex with unsplit DNA, it is reasonable to suppose that the formation of apoA-I complexes (P) with dimers (SS) leads to disordering of the secondary structure of dimer oligonucleotides and their dissociation into single-stranded oligonucleotides 1 and 2. Therefore, to describe the experimental data obtained we used a scheme of P–SS interaction similar to (4) but involving two stages [42, 43, 44]:

$$\begin{aligned}
&P + S \overset{K_1}{\Leftrightarrow} PS; \\
&PS + S \overset{K_2}{\Leftrightarrow} PS_2, \\
&\text{where } K_1 = [P]\cdot[S]/[PS], \quad K_2 = [PS]\cdot[S]/[PS_2], \\
&\Delta I_i(0) = I(0) - [P_o]\cdot J_P(0) - [S_o]\cdot J_S(0), \\
&\Delta J_i(0) = [PS]\cdot(J_{11}(0) - J_P(0) - J_S(0)) + [PS_2]\cdot(J_{12}(0) - J_P(0) - 2\cdot J_S(0)),
\end{aligned} \tag{5}$$

where $K_1$, $K_2$ are the dissociation constants for PS and $PS_2$ complexes at the two-stage interaction between P and S; $[P_o]$, $[S_o]$, $[P]$, $[S]$, $[PS]$, $[PS_2]$, $J_P(0)$, $J_S(0)$, $J_{11}(0)$, and $J_{12}(0)$ are the initial and equilibrium molar concentrations and intensities (at h = 0) of protein, apoA-I, oligonucleotides and protein complexes with single-stranded oligonucleotides (PS) and protein with dimer ($PS_2$) in each titration mixture, respectively; $I_i(0)$ is the intensity of scattering of the i-th titration mixture; $\Delta I_i(0)$ and $\Delta J_i(0)$ are the experimental and simulated values of differential intensities of the i-th mixture scattering (at h = 0).

Since dissociation constants of $P_1S_1$ complexes at the interaction of P with oligonucleotides coincide very closely (see Table 6), this allows us to make no difference in denotation of oligonucleotides 1 and 2 in scheme (5).

Optimization of criterion R was used to determine the values of dissociation constants K, $K_1$ and $K_2$ in models (4) and (5) for each set of experimental data presented in Table 6. The results obtained are listed in Table 7. One may see from Table 7 that in the presence of THC the experimental data are adequately described by the two-stage model (5). However, it is known that the values of equilibrium constants obtained by optimization for two stages ($K_1$ and $K_2$) are related to each other by statistical effects and the cooperation phenomenon [45, 46]: $K_1 = k/2$; $K_2 = 2 \cdot k/\chi$, where k is the equilibrium constant of dissociation, and $\chi$ is the cooperation coefficient. Then it can be shown that the values k = 19.84 μM ($k_{ass} = 5.04 \cdot 10^4$ $M^{-1}$) and $\chi = 0.24$ correspond to the values $K_1 = 9.92$ μM ($K_{ass1} = 1.01 \cdot 10^5$ $M^{-1}$) and $K_2 = 165.1$ μM ($K_{ass2} = 0.605 \cdot 10^4$ $M^{-1}$), which indicates the presence of negative cooperation ($\chi < 1$) [44] upon binding the single-stranded oligonucleotides to complex P. In the presence of cortisol, the data obtained are well described by the single-stage model (4) and much worse ($R_{min}$ = 25.2%) by model (5). This effect can be attributed to the fact that in this case, after the formation of complex PSS, SS dimer does not dissociate into single-stranded oligonucleotides but remains intact.

**Table 7. Values of equilibrium constants in models (4) and (5) and minimum values of criterion R obtained from experimental SAXS data listed in Table 6**

| Model (4): P + SS ⇔ PSS | | | | |
|---|---|---|---|---|
| THC | Cortisol | C, μM | | $R_{min}$, % |
| + | – | (not determined since $\Delta I_{1,2}(0) < 0$) | | – |
| – | + | 42.9 | | 2.1 |
| **Model (5): P + S ⇔ PS; PS + S ⇔ $PS_2$** | | | | |
| THC | Cortisol | $C_1$, μM | $C_2$, μM | $R_{min}$, % |
| + | – | 9.92 | 165.1 | 1.9 |
| – | + | 516.2 | 0.64 | 25.2 |

$J_{11}(0) = 2.923$ pulse · $s^{-1}$ · $\mu M^{-1}$;
$J_{12}(0) = 8.704$ pulse · $s^{-1}$ · $\mu M^{-1}$;
$J_P(0) = 1.68$ pulse · $s^{-1}$ · $\mu M^{-1}$;
$J_S(0) = 0.171$ pulse · $s^{-1}$ · $\mu M^{-1}$;
$J_{SS}(0) = 0.684$ pulse · $s^{-1}$ · $\mu M^{-1}$.

The presented data support our previous results [41, 42] showing the important role of THC but not other glucocorticoids in rupturing the secondary structure of DNA at its interaction with apoA-I, where apoA-I manifests specific enzymatic activity and THC acts as its cofactor.

Therefore, the results obtained using the SAXS method allow understanding the mechanism of enhancing the protein and DNA biosynthesis under the action of THC-apoA-I complex. An essential detail of this mechanism is the presence of reduced $\Delta^4$,3-ketogroup in tetrahydrocortisol A ring. The absence of such group in cortisol A ring prevents its involvement in the activation of both protein and DNA biosynthesis. Hence it follows that steroid regulation of gene expression should imply two independent mechanisms: the first one is the induction of gluconeogenesis enzymes, which is initiated through cytosol receptor of cortisol; the second mechanism consists in enhancement of protein and DNA biosynthesis under the action of reduced form of hormone. In this case, the target carrier of the latter is apolipoprotein A-I.

## Effect of THC-apoA-I Complex on DNA Interaction with RNA-Polymerase

Enhancement of gene expression under the action of steroid hormones implies the formation of conditions that facilitate DNA interaction with RNA-polymerase. In this connection, it was necessary to find out how the THC-apoA-I complex affects this mechanism. We attempted to answer this question also using the data of small-angle X-ray scattering.

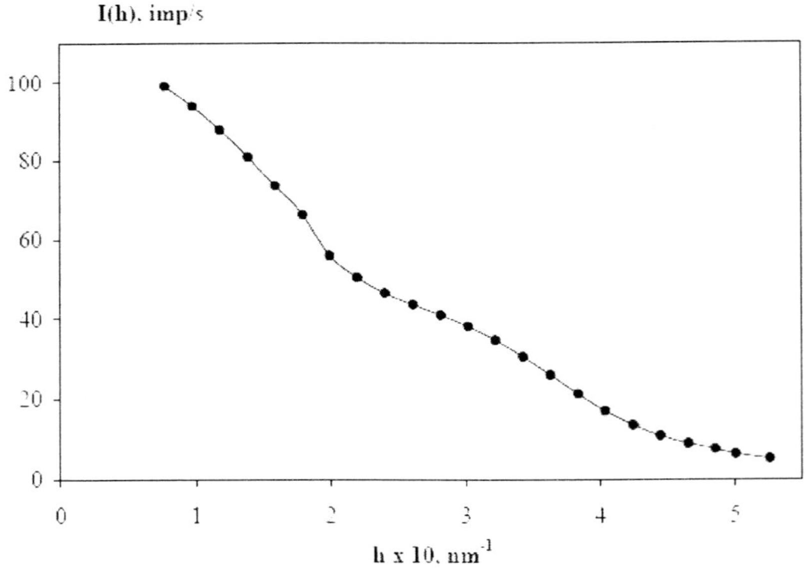


Figure 6. SAXS pattern of the RNA polymerase solution. The RNA polymerase concentration was 0.0416 mM. The patterns were corrected for the background scattering and collimation of the X-rays and smoothed.

Figure 6 shows the SAXS plot obtained for a solution of RNA-polymerase used in the study. Analysis of the obtained SAXS data shows that the preparations of RNA-polymerase, apoA-I and native DNA were fairly monodispersed. The structure of DNA-dependent RNA-polymerase from E. coli composed of 5 subunits ($\alpha_2\beta\beta'\sigma$) has been much studied by various methods, the methods of X-ray and neutron small-angle scattering among them [45, 46, 47]. However, the T7 RNA-polymerase used in our work is known to contain one subunit with molecular mass about 100 kDa [48]. The radius of gyration ($R_g$) value for RNA-polymerase T7 ($R_g$ = 6.83 ± 0.15 nm) we obtained from the SAXS data is close to the value ($R_g$ = 6.6 ± 0.3 nm) reported by the authors [45] for β′-subunit of RNA-polymerase from E. coli with a close molecular mass. The structural properties of apoA-I and native DNA from rat liver have been determined in our previous studies [41].

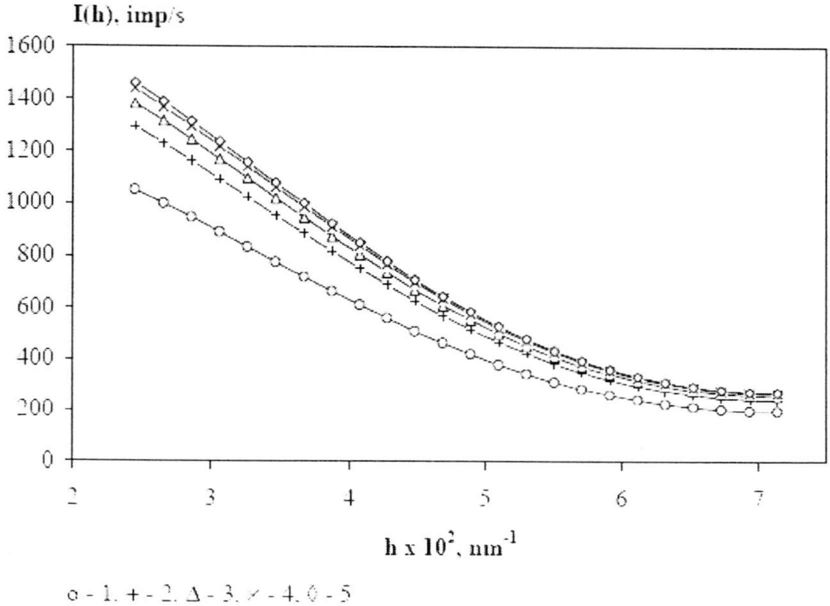


Figure 7. SAXS pattern of solutions of (*1*) free DNA and (*2-5*) DNA mixtures 1-4 (given in Table 2) with apoA-I and RNA polymerase in the presence of THC at 37°C.

The data of [41, 42] indicate the rupture of DNA secondary structure due to its interaction with THC-apoA-I complex and the formation of single-stranded regions. To estimate the interaction of RNA-polymerase and apoA-I with DNA by SAXS, the equilibrium mixtures of RNA-polymerase and DNA with apoA-I and its complexes with THC were used. Figure 7 presents the SAXS X-ray plots, Tables 8 and 9 list the values of structural characteristics of scattering particles obtained from DNA preparation and DNA mixtures with RNA-polymerase and apoA-I in the presence and in the absence of THC. The values of $R_g$ and I(0) for various mixtures of apoA-I with DNA and RNA-polymerase were calculated from experimental data (Tables 8, 9). One may see from Figure 8 and Tables 8, 9 that a pronounced interaction of DNA with apoA-I and RNA-polymerase was observed only in the presence of THC. The size of $P_nS$ complexes (here P stands for RNA-polymerase; S denotes the DNA aggregates with THC-apoA-I; n is the number of P molecules in the complex) within the accuracy of measurements corresponded to the size of DNA molecules in a free state ($R_g$ = 39.2 ± 0.6 nm), while reliably changed only the values of I(0) related to the concentration and molecular mass of $P_nS$ complexes, similar to [16, 49, 50]:

$$I(0) \sim N \cdot M^2, \qquad (1)$$

where N and M are respectively the number and molecular mass of the particles of $P_nS$ complexes involved into scattering. To evaluate the stoichiometry of $P_nS$ complex formed due to RNA-polymerase (P) interaction with DNA-apoA-I-THC complex (S), we used a simplified scheme [16, 41]:

$$n\,P + S \Leftrightarrow P_nS, \qquad (2)$$

where n is the number of protein molecules bound into $P_nS$ complexes. From estimates obtained in our SAXS experiments (Table 9) and from the data reported in [49, 50] one may assume that in these interactions the equilibrium is shifted essentially to the formation of complexes, hence, it is highly cooperative. Then, upon the formation of $P_nS$ complexes in the mixtures, parameter n in scheme (2) will grow up to some maximum value with increasing protein concentration ($C_p$) in the mixtures. Using (1) for estimation of the values of parameter n in each protein-DNA mixture, one may obtain the following expression [44, 49]:

$$n = M_S \cdot (T_{i0} - 1)/M_P, \qquad (3)$$

where $T_{i0} = (I_i(0)/I_0(0))^{1/2}$; $I_i(0)$ and $I_o(0)$ are the intensities of scattering to the zero angle (h = 0) from the i-th mixture and DNA initial preparation, respectively. Previously [19], using this method at the same component concentrations, we obtained the n value for apoA-I upon binding with DNA in the presence of THC equal to 54 ± 1. Table 9 presents the calculation results for parameter n and the estimates of $C_P/C_S$ balance for all RNA-polymerase mixtures with DNA used in the study in the presence of THC-apoA-I, taking $I_0(0)$ = 1605 cps. One may see that the n value in the mixtures does not exceed the values of the balance and in mixture 4 reaches its maximum at 6 ± 1, which corresponds to a 6:1 stoichiometry of the formed $P_nS$ complex upon interaction of RNA-polymerase with DNA molecules bound to THC-apoA-I complexes. RNA-polymerase of T7 phage is well known to conduct synthesis only from the double chain phage promoter with definite nucleotide sequence [48] and is widely used in different structural and functional investigations because the object is well studied and available, and has a simpler structure compared to eukaryotic polymerases. The data obtained with the SAXS method depend on the dispersion properties, that is, on the molecular weight and the shape of interacting molecules and their aggregates, so the monosubunit structure of T7 RNA-polymerase was the decisive factor in choosing the model system. Direct correlation between the action of the phage and eukaryotic RNA-polymerases on one and the same matrix was made in the study [51] using the RNA-polymerase of the SP6 phage and the yeast RNA-polymerase III as the examples. Both polymerases were shown to transcribe in a similar manner, for example, the histone octamer was transferred on approximately the same distances, which may account for the similarity of transcription initiation mechanisms. This points to some fundamental properties yet the transcription of eukaryotes is actually realized without phage polymerases. Research on the mechanisms of transcription with the participation of the RNA-polymerase II now presents a real challenge because of lack of an adequate experimental model [52].

**Table 9. Average values of the inertial radius $R_g$ and intensities of the X-ray scattering to the zero angle [I(0)], the number of RNA-polymerase molecules (n) in the complexes PnS[a] deduced from the SAXS data at various concentrations of RNA-polymerase ($C_p$) in the mixtures**

| Mixture | $C_p$, μM | $R_g$, nm | I(0), cps | n | $C_p/C_s$ |
|---|---|---|---|---|---|
| 1 | 0.140 | 39.0 ± 0.6 | 1764 ± 5[b] | 2 | 3 |
| 2 | 0.238 | 38.9 ± 0.7 | 1881 ± 7 | 4 | 5 |
| 3 | 0.380 | 39.2 ± 0.5 | 1960 ± 8 | 5 | 9 |
| 4 | 0.568 | 39.7 ± 0.8 | 1988 ± 8 | 6 | 13 |

a – the complexes of RNA- polymerase with DNA+[apoA-I-THC], where $C_s$=0.043 μM, concentrations: [apoA-I] = 3.55 μM and [THC] = 7.1 μM.
b – for the [DNA+apoA-I] complex (THC-free), I(0) = 1605 ± 9 cps

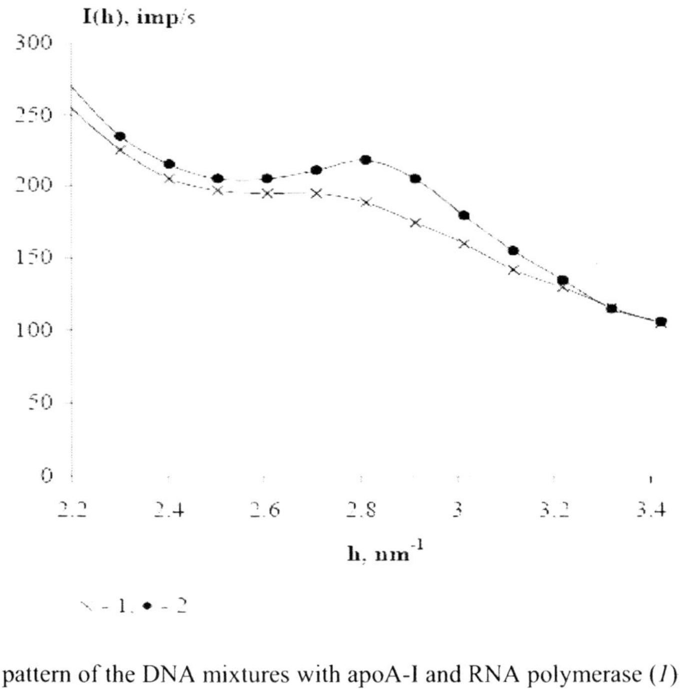


Figure 8. SAXS pattern of the DNA mixtures with apoA-I and RNA polymerase (*1*) in the absence and in the presence of THC (*2*). Concentrations: [DNA] – 0.043 μM; [apoA-I] – 3.55 μM; [THC] – 7.1 μM; [RNA polymerase] – 0.568 μM.

## Analysis of IR Spectra at the Interaction of Duplex $CC(GCC)_5{\cdot}GG(CGG)_5$ with Tetrahydrocortisol, Cortisol, and their Complexes with Apolipoprotein A-I

Dynamic characteristics and mechanisms of interaction of tetrahydrocortisol, cortisol, and their apolipoprotein A-I complexes with regulatory cis-elements of $CC(GCC)_5{\cdot}GG(CGG)_5$ type were studied using IR spectroscopy. The study allowed us to reveal the atomic groups from which starts their interaction with DNA and initiation of gene transcription.

*Interaction of Duplex with THC*

## Region of Nitrous Bases Absorption

Figure 9 shows IR spectra of duplex $CC(GCC)_5{\cdot}GG(CGG)_5$ for control samples and after its incubation with tetrahydrocortisol. The action of THC on the duplex shifted absorption bands of cytosine C=O bond from 1664 to 1656 $cm^{-1}$, a.b. of the intra-ring vibrations of C=N bond of guanine and cytosine, respectively, from 1573 to 1580 $cm^{-1}$ and from 1501 to 1505 $cm^{-1}$. An increase in the intensity of a.b. at 1492 $cm^{-1}$ (intra-ring vibrations of cytosine C=N bonds) was observed as well as increased splitting of the band at 1528 into 1526 and 1536 $cm^{-1}$. A change in intensity and the 1680 → 1684 $cm^{-1}$ shift was shown for the absorption band corresponding to guanine CO bond [53]. Increasing intensity and shifting the frequency of CO bond are caused by its participation in the formation of hydrogen bond between guanine and OH group of THC A ring, as well as by the duplex ordering.

The 1664 → 1656 $cm^{-1}$ shift of the absorption band indicates the formation of hydrogen bond between cytosine C=O group in the third position of THC A ring. Shifting of absorption bands at 1573 and 1501 $cm^{-1}$ (C=N bonds [53, 54]) is attributed to redistribution of electrons in aromatic bases caused by their disturbance with hormone and higher ordering. This is confirmed by the 1680 → 1684 $cm^{-1}$ shift. Note that in the duplex + THC mixture, in contrast to control sample, disappeared the band at 1688 $cm^{-1}$ assigned to a single-stranded structure. The results obtained give ground to conclude that a part of THC molecules interacts with cytosine, while another part interacts with the duplex guanine. This interaction involves C=O and C=N bonds of the both nitrous bases.

## Region of Ribose Phosphate Chain Absorption

Figure 9 shows a decrease in the frequency of absorption band with a broad peak from 1216-1240 to 1205 $cm^{-1}$. This is the P=O bond, whose asymmetric vibrations [53, 54, 55] evidence the formation of hydrogen bond between OH group of THC, most likely at $C_{17}$ carbon atom, and P=O bond of phosphate group of the duplex [56]. Shifting of this band was accompanied by a 1.6–1.8-fold increase in its intensity.

According to empirical rules of the H-bond formation and the data of quantum-chemical calculations performed with cytosine monophosphate [57], the observed changes in spectral parameters of absorption band assigned to phosphate group can be attributed to the formation of H-bond between water molecules and P-O groups. Frequency of the poly(C)poly(G) duplex at 1250 $cm^{-1}$ shifted by 5 $cm^{-1}$, while the band intensity increased no more than 1.4-fold [54]. The observed 11–35 $cm^{-1}$ shift in the P-O bond frequency and a 1.6–1.8-fold increase in intensity are attributed to the formation of the H-bond between P-O group of duplex and OH group of THC at $C_{17}$ carbon atom [56].

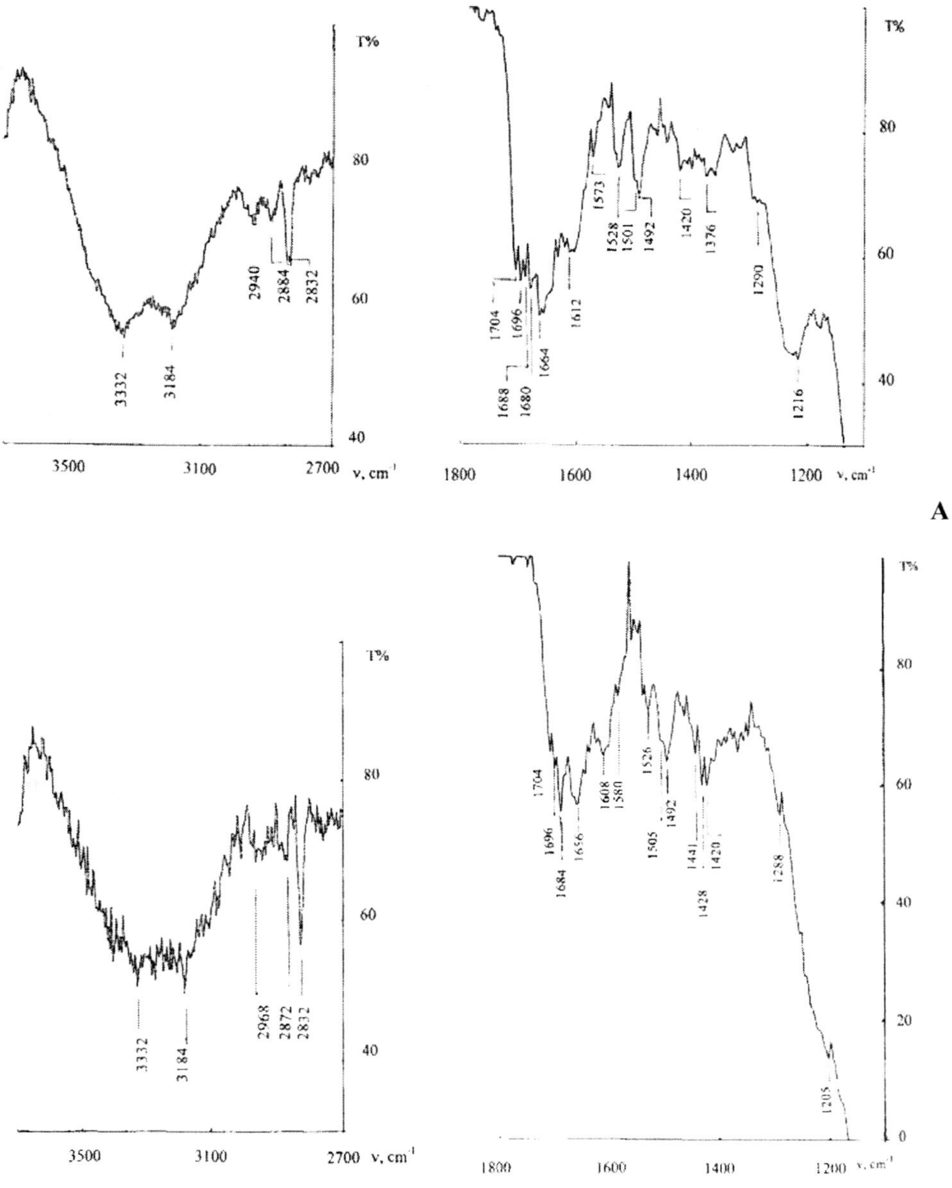


Figure 9. IR absorption spectra of duplex 5′-CC(GCC)$_5$ · GG(CGG)$_5$-3′Li$_2$ incubated without (A) or with (B) THC ($C_{THC} = 2{\cdot}10^{-6}$ M, $C_{buff} = 0.001$ M, pH = 7.2, relative humidity = 0%): ν = 2600-3700 $cm^{-1}$ (1), ν = 1000-1800 $cm^{-1}$ (2).

We revealed also the splitting and increase in intensity of the CH deformation vibration bands at 1420, 1428 and 1442 $cm^{-1}$. Note that in this case the ratio of peak intensities I $CH_{1420}$ / I $CO_{1650}$ comprises 0.8, while in all subsequent cases it does not exceed 0.1–0.2. In addition, an increase in intensity of absorption bands of CH stretching vibrations was recorded at 2832–2988 $cm^{-1}$ (triplet) and especially at 2832 $cm^{-1}$. The increased intensity, the shifted absorption bands of CH vibrations, and their splitting indicate changes in conformation of monosaccharides (desoxyribose) of the "boat" → "chair" type [56]. Variation in the parameters of CH vibrations is caused also by ordering of bases in the duplex after incubation with hormone.

However, the mechanism of duplex interaction with THC-apoA-I complex includes not only the common features but also some essential differences.

### *Interaction of Duplex with THC-apoA-I Complex*

### Region of Nitrous Bases Absorption

After the incubation of duplex with THC-apoA-I complex, changes were observed in some bands at 1500-1700 $cm^{-1}$, which either disappeared or varied in intensity and frequency (Figure 10). In particular, the band at 1696 $cm^{-1}$ disappeared. Absorption band at 1664 $cm^{-1}$ decreased in intensity, while the band at 1656 $cm^{-1}$ shifted to 1648 $cm^{-1}$. This shift can be attributed to the formation of additional hydrogen bond between $NH_2$ groups of apoA-I and carbonyl groups of the duplex cytosine. This interaction should be considered as the interaction of duplex with THC-apoA-I complex. An increase in intensity of absorption band at 1580 $cm^{-1}$ (intra-ring vibrations of guanine C=N bonds) and its shifting by 8 $cm^{-1}$ to the low-frequency region were observed. Absorption band at 1505 $cm^{-1}$ virtually disappeared; intensity of the band at 1492 $cm^{-1}$ (intra-ring vibrations of cytosine C=N bonds) decreased.

Disappearance of absorption band at 1696 and 1505 $cm^{-1}$ and decreased intensity and shifting of a.b. at 1580 and 1492 $cm^{-1}$ (intra-ring vibrations of C=N bonds of guanine and cytosine, respectively) indicate the dissociation (melting) of the duplex. A 1664 → 1648 $cm^{-1}$ frequency shift is attributed to the formation of H-bond between C=O group of cytosine and OH group of THC in complex with apoA-I, and evidences the duplex melting [54, 56].

### Region of Ribose Phosphate Chain Absorption

Absorption band of P=O bond shifted from 1205 to 1216 $cm^{-1}$ and decreased in intensity as compared to the case when only THC affected the duplex. Intensity of absorption bands at 1420 $cm^{-1}$ (CH deformation vibrations) and at 2828–2932 $cm^{-1}$ (CH stretching vibrations) decreased considerably. This is clearly seen from a.b. at 2828 $cm^{-1}$. Besides, these bands were shifted also. Splitting at 1420 $cm^{-1}$ disappeared (Table 10).

The 1205 → 1216 $cm^{-1}$ shift of absorption band and decrease in its intensity also indicate dissociation of the duplex [60]. This is attested also by decreasing intensity and disappearance of splitting of the bands at 1420 and 2828–2932 $cm^{-1}$ [58].

Shifting the absorption bands of C=O bonds led also to a 8–12 $cm^{-1}$ shift of π-conjugated absorption bands of NH stretching vibrations (3332 → 3344 and 3184 → 3192 $cm^{-1}$) to the high-frequency region. An additional maximum appeared at 3276 $cm^{-1}$ associated with the presence of apoA-I (Figure 10).

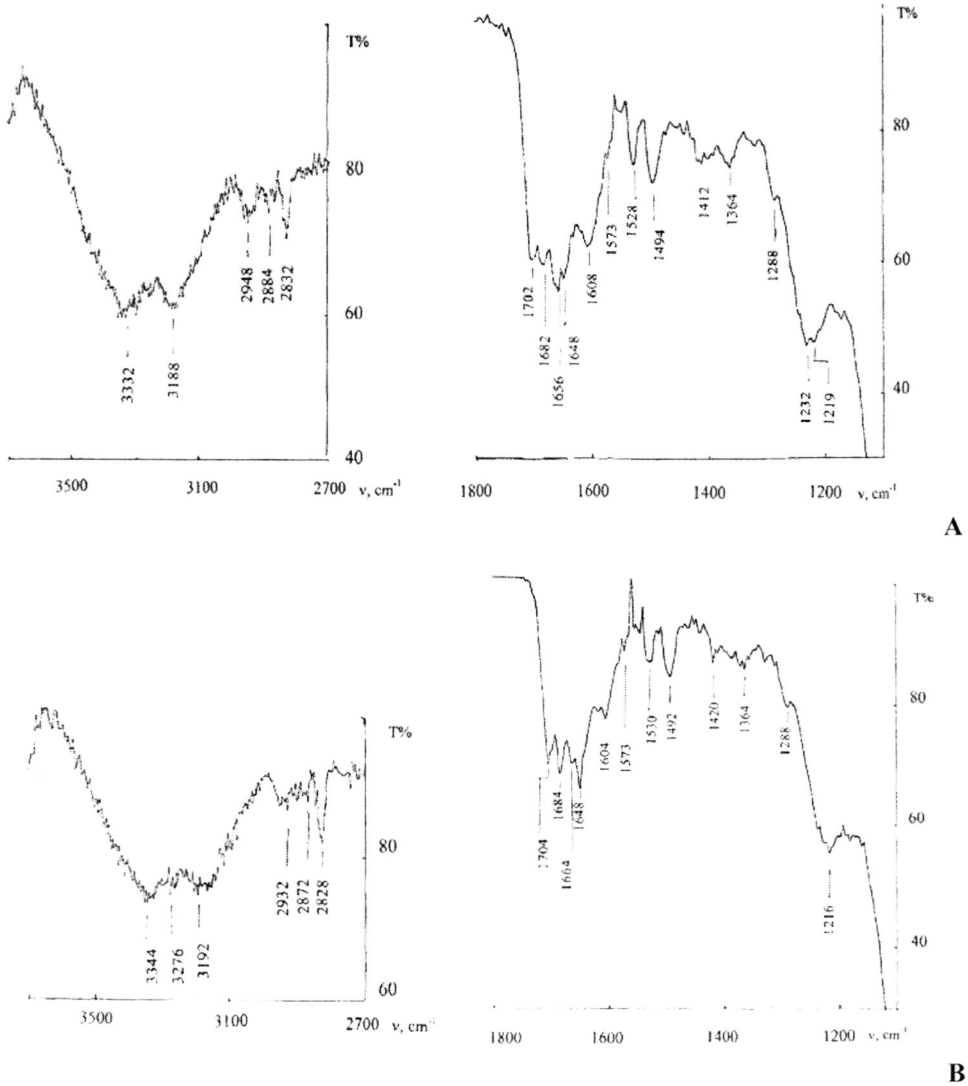


Figure 10. IR absorption spectra of duplex 5′-CC(GCC)$_5$ · GG(CGG)$_5$-3′Li$_2$ incubated with apoA-I (A) or THC-apoA-I complex (B). ($C_{THC}$ = $2 \cdot 10^{-6}$ M, $C_{apoA\text{-}I}$ = $2 \cdot 10^{-6}$ M, $C_{buff}$ = 0.001 M, pH = 7.2, relative humidity = 0%): ν = 2600-3700 $cm^{-1}$ (1), ν = 1000-1800 $cm^{-1}$ (2); 5′-CC(GCC)$_5$ · GG(CGG)$_5$-3′+ apoA-I (control) (A); 5′-CC(GCC)$_5$·GG(CGG)$_5$-3′+THC-apoA-I (B).

### *Mechanism of Interaction of Duplex with Apolipoprotein apoA-I*

Figure 10 shows the disappearance of band at 1696 $cm^{-1}$ and band shifts: 1704 → 1702, 1686 → 1682, and 1664 → 1656 $cm^{-1}$. Note the occurrence of absorption band at 1648 $cm^{-1}$ in the spectrum. In addition, the band at 1573 $cm^{-1}$ decreased in intensity (intra-ring vibrations of guanine C=N bonds) [54]. Absorption band at 1501 $cm^{-1}$ disappeared, a.b. shift was observed at 1492 → 1494 $cm^{-1}$ accompanied by decrease in intensity (intra-ring vibrations of cytosine C=N bonds) [54].

Intensity of vibrations decreased at 1420 $cm^{-1}$ (CH deformation vibrations), splitting in this region disappeared (Table 10). In the region of stretching vibrations of CH bonds, a band shift (2940 → 2948 $cm^{-1}$) and decrease in intensity of band at 2832 $cm^{-1}$ were observed, which indicates a partial disordering of monosaccharides in the duplex. The 3184 → 3188 $cm^{-1}$ shift of absorption band was observed (NH stretching vibrations). The band maximum at 1224 $cm^{-1}$ split into two peaks at 1219 and 1232 $cm^{-1}$ (P=O bond) and decreased in intensity. Such splitting of absorption band and decrease in intensity evidence the formation of hydrogen bond between apoA-I and duplex.

Disappearance of absorption bands at 1696 and 1501 $cm^{-1}$, band shifts 1704 → 1702, 1686 → 1682, and finally 1664 → 1648, as well as decreased intensity and disappearance of splitting in the region of 1420 $cm^{-1}$ indicate the destruction of helical structure of duplex under the action of apoA-I [54, 56].

Therefore, the cooperative action of THC and apoA-I in a single complex is aimed at attaining the additive result that consists in despiralization of DNA and rupture of the H-bond between GC pairs. Probably, local melting of DNA may further include the hydrophobic interactions between nitrous bases and hydrophobic segments of alphahelices of apolipoprotein A-I. This increases the extent of DNA melting and forms a basis for its interaction with RNA-polymerase.

### *Mechanism of Interaction of Duplex with Cortisol*

Analysis of the spectra of interaction between duplex and cortisol revealed band shifts: 1688 → 1698, 1654 → 1662, 1614 → 1608, 1536 → 1524, and 1494 → 1492 $cm^{-1}$. The 1688 → 1698 $cm^{-1}$ band shift indicates an increased ordering of the duplex under the action of cortisol [53]. The high-frequency band shift 1654 → 1662 $cm^{-1}$ can be attributed to the formation of hydrogen bonds between C=O group of the hormone A ring and NH group involved in the formation of internal H-bond of GC pairs. Since CO and NH bonds of bases are π-conjugated, a change in frequency of NH bond changes also the CO bond frequency. The bond shift 1494 → 1492 $cm^{-1}$ is caused by disturbance of C=N bonds of cytosine intra-ring vibrations.

A shift of the NH–CO bond to a high-frequency region was actually found. It comprises 35–40 $cm^{-1}$. Redistribution of intensities of CH stretching vibration absorption bands was also found. Two equivalent bands at 2778 and 2924 $cm^{-1}$ are seen in the spectrum of duplex, while in the spectrum of duplex + cortisol the intensities of corresponding bands at 2752 and 2950 $cm^{-1}$ differ by a factor of 3–4. It means that in the first case, monosaccharides of the duplex contain two conformations, "boat" and "chair", approximately in equal amounts, while in the second case, the "chair" conformation prevails, being energetically more advantageous [56].

Absorption band at 1224 $cm^{-1}$ of the duplex P=O bond splits into 1238 and 1220 $cm^{-1}$ bands upon interaction of duplex with cortisol. This splitting is attributed to the formation of hydrogen bond between P=O and OH group of hormone, probably, in $C_{17}$ position [56]. However, this bond is weaker than that formed at the interaction of duplex with THC, since both the shift and the intensity increment of P=O bond are weaker.

*Mechanism of Interaction of Duplex with Cortisol-apoA-I Complex*

## Region of Nitrous Bases Absorption

Analysis of the spectra of interaction between duplex and cortisol-apoA-I complex showed that some bands in the region of nitrous bases absorption (1500–1700 $cm^{-1}$) changed their intensity and frequency. In particular, absorption band at 1698 $cm^{-1}$ significantly decreased its intensity relative to the band at 1664 $cm^{-1}$; on the other hand, absorption band at 1664 $cm^{-1}$ was shifted by 2 $cm^{-1}$ to the short-wave region with respect to the control spectrum (duplex + cortisol). A maximum appeared at 1648 $cm^{-1}$, its intensity being lower than that at 1664 $cm^{-1}$; however, its is seen quite clearly. Absorption band at 1614 $cm^{-1}$ shifted to the long-wave region to 1604 $cm^{-1}$ (C=C bond) [59]. Redistribution of intensities and band shifts were found in the region of 1536 and 1492 $cm^{-1}$. Absorption band at 1492 $cm^{-1}$ shifted to 1496 $cm^{-1}$ and decreased in intensity, while a.b. at 1524 shifted to 1528 $cm^{1}$.

**Table 10. Frequency characteristics of stretching and deformation vibrations of bonds in $CC(GCC)_5 \cdot GG(CGG)_5$ duplex before and after interaction with THC, cortisol and their apoA-I complexes**

| No. | Compound | $\nu_{CO}$ $cm^{-1}$ | $\nu_{NH}$ $cm^{-1}$ | $\nu_{CN}$ $cm^{-1}$ | $\nu_{CH\ str}$ $cm^{-1}$ | $\nu_{CH\ def}$ $cm^{-1}$ | $\nu_{PO2}$ $cm^{-1}$ |
|---|---|---|---|---|---|---|---|
| 1 | Duplex | 1704<br>1696<br>1688<br>1680<br>1664 | 3332<br>3184 | 1573<br>1501<br>1492 | 2940<br>2884<br>2832 | 1442 weak<br>1420 | 1290<br>1224 |
| 2 | Duplex + THC | 1704<br>1696<br>1684<br>1656 | 3332<br>3184 | 1580<br>1505<br>1492 | 2968<br>2872<br>2832↑ | 1441<br>1428<br>1420 | 1288<br>1205↑ |
| 3 | Duplex + apoA-I + THC | 1704<br>1684<br>1648 | 3344<br>3276<br>3192 | 1573<br>1492↓ | 2932<br>2872<br>2828↓ | 1420↓<br>no splitting | 1288<br>1216 |
| 4 | Duplex + cortisol | 1696<br>1658 | 3348<br>3204 | 1494 | 2944<br>2884<br>2832↑ | 1428<br>1404 | 1290<br>1224 |
| 5 | Duplex + apoA-I + cortisol | 1698↓<br>1683↓<br>1664↑<br>1648 | 3320<br>3280<br>3170 | 1574<br>1496↓ | 2932<br>2880<br>2832↓ | 1420<br>no band | 1288<br>1214 |
| 6 | Duplex + apoA-I | 1702<br>1682<br>1656<br>1648 | 3332<br>3260<br>3188 | 1573↓<br>1494↓ | 2948<br>2884<br>2832↓ | 1412↓<br>no splitting | 1288<br>1232<br>1219 |

Note: Frequency measurement error was 0.5 cm–1 for 1800-1000 cm–1 region, and 1 cm–1 for 2700-3600 cm–1 region. Symbols (↑↓) denote increase or decrease in intensity.

## Region of Ribose Phosphate Chain Absorption

Absorption band of P=O bond shifted from 1220 to 1214 $cm^{-1}$ and became comparable in frequency and intensity with the band at 1216 $cm^{-1}$ in the spectrum of duplex + THC-apoA-I. Intensity of bands at 1082 and 1054 $cm^{-1}$ differed from that in the duplex and was somewhat lower as compared to the spectrum of duplex + cortisol. In addition, these bands shifted from 1077 to 1082 $cm^{-1}$. Bands in the region of 1400–1470 $cm^{-1}$ virtually disappeared, while in the duplex and in the duplex + cortisol complex they were observed. From the analysis of absorption bands it follows that local melting occurs also in this case, but it does not lead to complete dissociation of the duplex into single-stranded oligonucleotides. Most important results of changes that occur in the IR spectra upon interaction of duplex with THC, cortisol and their apoA-I complexes are presented in the Table 8.

**Table 8. Average values of the inertial radius $R_g$ and intensities of the X-ray scattering to the zero angle [I(0)] obtained from SAXS data for mixtures of RNA-polymerase[a] and DNA[b] in the presence (+) and absence (-) of THC[c] and apoA-I[d]**

| THC | apoA-I | RNA-polymerase | $R_g$, nm | I(0), cps |
|---|---|---|---|---|
| - | - | - | 39.1 ± 0.5 | 1450 ± 11 |
| - | - | + | 38.8 ± 0.7 | 1485 ± 15 |
| - | + | + | 39.3 ± 0.6 | 1490 ± 12 |
| + | + | - | 39.2 ± 0.4 | 1605 ± 10 |

Concentrations: a – 0.140 μM, b – 0.043 μM, c – 7.1 μM, d – 3.55 μM.

Therefore, study on the mechanism of interaction of $CC(GCC)_5{\cdot}GG(CGG)_5Li_2$ duplex with THC, cortisol and THC-apoA-I complex indicated the formation of hydrogen bonds between OH group of THC A ring and cytosine C=O group, and with cytosine NH bond in the case of cortisol. In addition, THC and cortisol form hydrogen bonds with $PO_2$ group of the duplex desoxyribose phosphate and with OH group of monosaccharide. The interaction of apoA-I with the duplex is accompanied by the formation of hydrogen bonds with cytosine C=O group and P=O group of desoxyribose phosphate. The interaction of THC and apoA-I with the duplex occurs at the same active point, namely, with C=O group of cytosine. Structural transition of the order → disorder type takes place in the duplex under the action of complexes THC-apoA-I, cortisol-apoA-I or apoA-I.

# Discussion

The results of incubation of hepatocytes with steroid hormones, apoA-I and their complexes demonstrated that cortisol and tetrahydrocortisol did not enhance the protein and DNA biosynthesis rate. The cortisol-apoA-I complex inhibited the protein and DNA biosynthesis in hepatocytes. On the contrary, the THC-apoA-I complex largely enhanced the protein and DNA synthesis in these cells. The results obtained indicate that reduced forms of steroid hormones (tetrahydrocortisol) are biologically active and regulate gene expression in liver parenchymal cells.

A significant acceleration of protein and DNA biosynthesis under the action of this complex allows an assumption that the described phenomenon underlies the mechanisms of regeneration of parenchymal cells in organs and tissues. Previously it was shown that these processes are initiated by resident macrophages, which possess a remarkable ability to reduce steroid hormones yielding tetrahydrocompounds. Now we know that they contribute also to the formation of biologically active complex THC-apoA-I, which enhances the synthesis of DNA, RNA and protein in parenchymal cells [17, 42].

It is known that only the macrophages show a pronounced 5α- and 5β-reductase activity, which determines the reduction of $\Delta^4$,3-ketogroup of steroid hormones and formation of tetrahydroforms [60]. Via phagocytosis of the products of cell degradation, macrophages cooperatively absorb $HDL_3$ and steroid hormones. The same process occurs when macrophages are stimulated by bacterial lipopolysaccharides [17]. In secondary lysosomes, $HDL_3$ are disintegrated, which is accompanied by apoA-I release and reduction of steroid hormones. Then the THC-apoA-I complex forms, which is resecreted to interstitial space. Further the complex is captured by the cells (hepatocytes) and transferred to the nucleus, where it interacts with deproteinized DNA regions [41, 42].

The mechanism of this interaction is shown in the scheme (Figure 11).

The interaction mechanism was studied in more detail with DNA isolated from rat liver and with synthetic oligonucleotides, which are the assumed sites of interaction between the complex and DNA. Among the 54 regions of interaction between THC-apoA-I complex and DNA revealed by the SAXS method [41, 42], there were only 1-2 cases when the single-stranded DNA chains split by S1 nuclease formed. Probably, binding of RNA-polymerase is also partially caused by unspecific interaction. It seems likely that a fraction of RNA-polymerase was bound to DNA independently of newly formed single-stranded (S1-nuclease-sensitive) fragments. This is not surprising, since RNA-polymerase can bind also to the native DNA [48]. It is significant that in this case the partially untwisted sections of DNA double chain are the targets for RNA-polymerase, i.e., the specific change we have found in the DNA secondary structure under the action of THC-apoA-I initiates cooperative binding of RNA-polymerase. It was shown [61] that the formation of specific elongational complex of transcription with RNA-polymerase II in vitro may go without promoter-depending initiation with the availability of 3`-extending ends from which initiation is being realized. The pattern we suggest is characterized by the transcription initiation in vivo possibly taking place on the newly appearing internal single-stranded regions of DNA under the conditions of regeneration at some adaptive processes.

The reduced form of glucocorticoid, tetrahydrocompound, was shown to be the biologically active form of hormone involved in the enhancement of gene expression composing THC-apoA-I complex. In this case, apoA-I acts as a targeted carrier of hormone, and the reduced form of hormone facilitates the rupture of hydrogen bonds between the pairs of DNA nitrous bases, causing its local melting and creating prerequisites for RNA-polymerase interaction with single-stranded regions of DNA. The presented mechanism of gene expression enhancement may be the reason for activation of DNA, RNA and protein biosynthesis in hepatocytes under the action of THC-apoA-I complex [33, 41]. In this case, the biological function of THC-apoA-I complex is similar to the function of some transcriptional factors changing the DNA structure [62]. We supposed that the most probable sites of interaction with THC-apoA-I complex are the GC-rich DNA regions.

Replicas of (GCC)·$(CGG)_n$ type are very frequent in both prokaryotic and eukaryotic genomes. Database search we performed over the GenBank EMBL DNA library showed the presence of these sequences in all genome structures: exons, introns, 5′- and 3′-nontranslated regions, and very often in promotor regions of genes. The number of such replicas varies in a broad range. For instance, a sequence of $CC(GCC)_3$ type was found in human apolipoprotein A-I gene. This sequence locates at the position of +6...+16 nucleotide pairs from the translational stop codon [56]. $(GCC)_{14}$ sequence was found in 5′-flanking region of the gene responsible for synthesis of the human multidrug-resistance protein [31]. Similar sequences (with a varying number of replicas) were detected in the gene accounting for the synthesis of human pyruvate dehydrogenase kinase [63], in the gene controlling the synthesis of human fibroblast growth factor receptor [64], etc.

The regions of (GCC)·$(CGG)_n$ replicas in gene promotors contain binding sites of some transcriptional factors, which indicates their important role in the regulation of gene expression. For example, among transcriptional factors capable of binding to $(GCC)_n$ replica in promotor region of the human multidrug-resistance gene are such factors as SpI, Adf-I, EGR2 and others (GenBank).

We were the first to demonstrate that tetrahydrocortisol-apolipoprotein A-I complex can actively bind to the replicas of (GCC)·$(CGG)_n$ type. Interaction of the complex with DNA results in local melting of the latter. This allows RNA-polymerase to sit on DNA and initiate the mRNA synthesis. Previously it was shown that this mechanism controls also the formation of ribosomal RNA [17] (see the Scheme).

The process of DNA melting in the region of THC-apoA-I binding sites was simulated with the duplex of $CC(GCC)_5$·$GG(CGG)_5$ type. The interaction of THC-apoA-I complex with SS duplex was shown to cause the formation of a more composite complex, which dissociates into mutually complementary oligonucleotides. They in turn interact with free P molecules to form new PS complexes.

IR-spectroscopy study on the mechanism of interaction of DNA and duplex $CC(GCC)_5 \cdot GG(CGG)_5$ with THC or THC-apoA-I complex revealed the formation of hydrogen bonds between OH group of the THC A-ring and C=O group of cytosine or guanine. This interaction is of a competitive nature; it is accompanied by the rupture of hydrogen bonds between C=O and –NH groups in GC pairs. Local melting of DNA occurs at the sites of its interaction with THC-apoA-I complex. The initiating role here belongs to the hormone. However, the mechanism of DNA melting may further include the hydrophobic interactions between the released nitrous bases and hydrophobic segments of amphipathic alphahelices of apolipoprotein A-I. This creates prerequisites for the DNA interaction with RNA-polymerase. In this way the regulation of gene expression proceeds with the participation of reduced forms of steroid hormones.

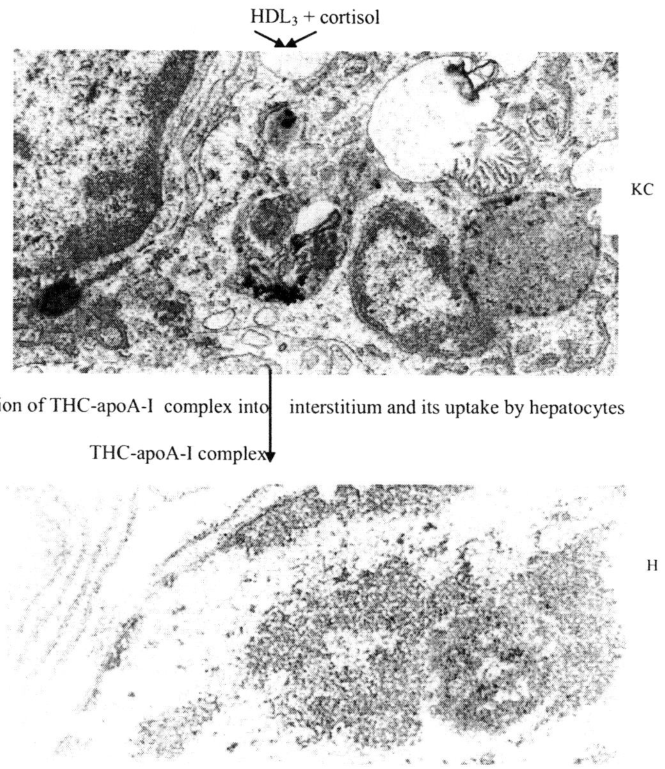


Figure 11. The role of Kupffer cells in enhancement of gene expression and rate of protein biosynthesis in hepatocytes (scheme): 1. cooperative uptake of $HDL_3$ and cortisol, reduce of it, and form of biological active complex – tetrahydrocortisol-apolipoprotein A-I (THC-apoA-I); 2. secretion of THC-apoA-I complex by Kupffer cells into interstitium and its uptake by hepatocytes; 3. activation of nucleolar DNA, enhancement of gene expression and rate of protein biosynthesis. A - Kupffer cell. B – Hepatocyte. Electron microscopy. Scale 0.5 mkm

# References

[1] Favre G., Blancy E., Tournier G.F., Soula G. (1989). Proliferative effect of high density lipoprotein (HDL) and HDL fractions ($HDL_1$, $HDL_3$) on virus transformed lymphoblastoid cells. *Biochim. Biophys. Acta.* V. 1013, p. 118-124.

[2] Biran S., Horowitz A.T., Fuks Z., Vlodavsky G. (1983). HDL and extracellular matrix promote growth and plating efficiency of normal mammary cells in serum. *Int. J. Cancer.* V. 31, p. 557-566.

[3] Tamagaki T., Sawada S., Imamara H., Tada Y., Yamasaki S., Taratani A., Sato T., Komatsu S., Akamatsu N., Yamagami M., Kobayashi K., Kato K., Yamamoto K., Shiraki K., Yamada K. (1996). Effects of high-density lipoproteins on intercellular pH and proliferation of human vascular endothelial cells. *Atherosclerosis.* V. 123, p. 73-82.

[4] Resnik T.I., Bochkov V.N., Hahn A.W.A., Philippova M.P., Buhler F.R., Tkachuk V.A. (1995). Low and high-density lipoproteins as mitogenic factor for vascular smooth muscle cells. Individual, additive, and synergistic effects. *J. Vasc. Res.* V. 32, p. 328-338.

[5] Favre G., Tazi K.A., Le Gaillard F., Bennis F., Hachem H., Soula G. (1993). High-density $lipoprotein_3$ binding sites are related to DNA biosynthesis in the adenocarcinoma cell like A549. *J. Lipid Res.* V. 34, p. 1093-1106.

[6] Desanctis J.B., Blanca I., Bianko N.E. (1995). Effect of different lipoproteins on the proliferative response of interleukin-2-activated T lymphocytes and large granular lymphocytes. *Clin. Sci.* V. 89, N 5, p. 511-519.

[7] Desanctis J.B., Blanca I., Bianko N.E. (1997). Secretion of cytokines by natural killer cells primed with interleukines-2 and stimulated with different lipoproteins. *Immunology.* V. 90, p. 526-533.

[8] Ashby D.T., Rye K.A., Clay M.A., Vadas M.A., Gamble J.R., Barter P.J. (1998). Factors influencing the ability of HDL to inhibit expression of vascular cell adhesion molecule-1 in endothelial cells. *Arterioscler. Thromb. Vasc. Biol.* V. 18, N 9, p. 1450-1456.

[9] Jonas A., Steinmet A., Churgay L. (1993). The number of amphipathic alphahelical segments of apolipoprotein A-I, E, and A-IV determines the size and functional properties of their reconstituted lipoprotein particles. *J. Biol. Chem.* V. 268, p. 1596.

[10] Handwerger S., Myers S., Richards R., Richardson B., Turzai L., Moeykins C., Meyer T., Anantharamahian G.M. (1995) Apolipoprotein A-I stimulates placental lactogen expression by human trophoblast cell. *J. Endocrinol.* V. 136, p. 5555-5560.

[11] Blackburn W.D., Dohlman J.D., Venkatachalapathi Y.V. (1991). Apolipoprotein A-I decreases neutrophil degranulation and superoxide production. *J. Lipid. Res.* V. 32, p. 1911-1918.

[12] Panin L.E., Biushkina N.G., Polyakov L.M. (1992). Quantitative characteristics of interactions of blood lipoproteins and steroid hormones. *Bull. Exp. Biol. Med.* (Russian). V. 114, N 7, p. 34-36.

[13] Panin L.E. (1998). Phenomenon of protein biosynthesis stimulation in parenchymal cells of organs and tissues by resident macrophages. *Bulletin Siberian Division of RAMS.* N 3, p. 11-23.

[14] Panin L.E., Russkikh G.S., Polyakov L.M. (2000). Detection of apolipoprotein A-I, B, and E immunoreactivity in the nuclei of various rat tissue cells. *Biochemistry* (Moscow). V. 65, p. 1419-1423.

[15] Panin L.E., Polyakov L.M., Usynin I.F., Kuzmenko A.P., Poteryaeva O.N. (1991). Immunochemical identification of apolipoprotein A-I in chromatin of rat hepatocyte nuclei. *Biopolymers and Cell* (Russia). V. 8, N 2., p. 44-49.

[16] Panin L.E., Tuzikov F.V., Tuzikova N.A., Khar'kovskii A.V., Usynin I.F. (1999). The influence of tetrahydrocortisol-apolipoprotein A-I complex on the hepatocyte's protein synthesis and on DNA secondary structure. *Molecular Biology* (Russia). V. 33, p. 1-6.

[17] Panin L.E., Maksimov F.V., Usynin I.F., Korostyshevskaya I.M. (2002). Activation of nucleolar DNA expression in hepatocytes by glucocorticoids and high density lipoproteins. *J. Steroid Biochem. and Mol. Biol.* V. 81, p. 69-76.

[18] Panin L.E., Svechnikova I.G., Mayanskaya N.N. (1995). Cooperative effect of the high density lipoproteins and adaptive hormones in a chromatin activation of rat's liver. *Ukrainian Biochem. J.* (Russia). V. 67, p. 64-70.

[19] Panin L.E., Gimautdinova O.I., Polyakov L.M., Nayakshina T.N. (1998). Peculiarities of the interaction of cortisol, tetrahydrocortisol and its complexes with apolipoprotein A-I with eukaryotic DNA. *Molecular Biology* (Russia). V. 32, p. 447-451.

[20] Yudaev N.A., Afinogenova S.A., Krekhova M.A. (1976). *Biochemistry of Hormones and Hormonal Regulation.* (Russia), Moscow, Nauka, p. 171-227.

[21] Vogt V.M. (1980). Purification and properties of S1nuclease from Aspergillus. *Method Enzymol.*, V. 65, p. 248-255.

[22] Maniatis T., Fritsch E., Sambrook J. (1982). *Molecular Cloning: A Laboratory Manual.* Cold Spring Harbor, NY: Cold Spring Harbor Lab. Cited by *Methods of Genetic Engineering. Molecular Cloning.* (Russia, 1984), Moscow, Mir, p. 272.

[23] Hatch F.T., Lees R.S. (1968). Practical methods for plasma lipoprotein analysis. *Advanc. Lipid. Res.* V. 6, p. 2-68.

[24] Lowry O.H., Rosebrough N.J., Farr A.L., Randall R.J. (1951). Protein measurement with the Folin phenol reagent. *J. Biol. Chem.* V. 193, p. 265-275.

[25] Seglen P.O. (1976). Preparation of isolated rat liver cells. *Meth. Cell. Biol.* V. 3, p. 29-83.

[26] Usynin I.F., Panin L.E., Kharkovskii A.V., Goryacheva M.V. (1995). Role of non-parenchymal liver cells in activation of protein synthesis in hepatocytes on stimulation of the mononuclear phagocyte system by lipopolysaccharide. *Biochemistry* (Moscow). V. 60, N 4, p. 399-405.

[27] Dicker P., Rozengurt E. (1980). Phorbol esters and vasopressin stimulate DNA synthesis by a common mechanism. *Nature.* V. 287, p. 607-612.

[28] Weigand K., Otto J., Schopt R. (1974). Ficoll density separation of enzymatically isolated rat liver cells. *Acta Hepato-Gastroenterol.* V. 21, p. 245-253.

[29] Karlinsey J., Stamatoyannaopoulos G., Enver N. (1989). Simultaneous purification of DNA and RNA from small numbers of eukaryotic cells. *Anal. Biochem.* V. 180, p. 303-306.

[30] Panin L.E., Tuzikov F.V., Tuzikova N.A., Gimautdinova O.I., Polyakov L.M. (2002). Interaction of tetrahydrocortisol-apolipoprotein A-I complex with eukaryotic DNA and with single-stranded oligonucleotides. *Molecular Biology* (Russia). V. 36, N 1, p. 96-102.

[31] Grant C.E., Kurz E.U., Cole S.P.C., Deeley R.G. (1997). Analysis of the intron-exon organization of the human multidrug-resistance protein gene (MRP) and alternative splicing of its mRNA. *Genomics*. V. 45, N 2, p. 368-379.

[32] Svergun D.I., Feygin L.A. (1986). *X-Ray and Neutron Small-Angle Scattering*. Moscow, Nauka, p. 50-279.

[33] Panin L.E., Tuzikov F.V., Tuzikova N.A., Usynin I.F., Gimautdinova O.I. (2001). The effect of tetrahydrocortisol-apolipoprotein A-I complex on the RNA polymerase interaction with eukaryotic DNA and the rate of protein biosynthesis in hepatocytes. *Bioorganic Chemistry* (Russia). V. 27, N 2, p. 114-119.

[34] Kunitsyn V.G., Panin L.E., Polyakov L.M. (2001). Anomalous change of viscosity and conductivity in blood plasma lipoproteins in the physiological temperature range. *Int. J. Quantum Chem*. V. 81, p. 348-369.

[35] Sergeev P.V., Galenko-Yaroshevsky P.A., Shimanovsky N.L. (1996). *Essays on Biochemical Pharmacology*. Moscow, Pharmedinfo, p. 270.

[36] Romanenko E.B., Demidenko Z.N., Vanyushin B.F. (1998). Activities of RNA-polymerase, DNA-polymerase, DNA-methyltransferase and sfingomielase in the nuclei of different age rat's liver. *Biochemistry* (Russia). V. 63, p. 194-199.

[37] Yenukashvili N.I., Lobov I.B., Podgornaya O.I. (1999). Human nuclear matrix contains a protein complex specifically binding alpha-satellite DNA in vitro. *Biochemistry* (Russia). V. 64, N 4, p. 564-573.

[38] Lobov I.B., Podgornaya O.I. (1999). The role of nuclear matrix proteins in the formation of heterochromatin. *Cytology* (Russia). V. 41, N 7, p. 562-573.

[39] Svetlova E.Yu., Rasin S.V., Filipsky Ya. (1997). Mapping of S1-nuclease-hypersensitive regions in the III chromosome of saccharomyces cerevisiae yeast. *Doklady AN* (Russia). V. 353, N 6, p. 831-833.

[40] Boulikas T. (1995). Chromatin domains and prediction of MAR sequences. *Int. Review of Cytology*. V. 162, p. 279-388.

[41] Panin L.E., Tuzikov F.V., Tuzikova N.A., Gimautdinova O.I., Polyakov L.M. (2000). Interaction of eukaryotic DNA with apolipoprotein A-I and its complexes with glucocorticoid. *Biophysics*. V. 45, N 4, p. 611-619.

[42] Panin L.E., Gimautdinova O.I., Kuznetsov P.A., Akimzhanova M.V., Tuzikov F.V. (2002). The effect of tetrahydrocortisol-apolipoprotein A-I complex on the secondary structure of eukaryotic DNA and its interaction with RNA-polymerase. *Biochemistry* (Russia). V. 67, N 7, p. 953-958.

[43] Tuzikov F.V., Zinoviev V.V., Malygin E.G., Ankilova V.N., Moor N.A., Lavrik O.I. (1988). Application of the small-angle X-ray scattering technique for the study of equilibrium enzyme-substrate interactions of phenylalanyl-tRNA synthetase from E. coli with $tRNA^{Phe}$. *FEBS Lett*. V. 232, p. 107-110.

[44] Tuzikov F.V., Zinoviev V.V., Vavilin V.I., Malygin E.G. (1996). Application of the small-angle X-ray scattering technique for the study of two-step equilibrium enzyme-substrate interactions. *Biopolymers*. V. 38, p. 131-139.

[45] Stockel P., May R.P., Strell I., Cejka Z., Hoppe W., Heumann H. (1980). The core subunit structure in RNA polymerase holoenzyme determined by neutron small-angle scattering. *Eur. I. Biochem*. V. 112, p. 411-417.

[46] Meisenberger O., Heumann H., Pilz I. (1981). Small-angle X-ray study of DNA-dependent RNA polymerase holoenzyme from Escherichia coli. *FEBS Lett.* V. 123, p. 22-24.

[47] Heumann H., Lederer H., Baer G., May R.P., Kjems J.K., Crespi H.L. (1988). Spatial arrangement of DNA-dependent RNA polymerase of Escherichia coli and DNA in the specific complex. A neutron small angle scattering study. *J. Mol. Biol.* V. 201, p. 115-125.

[48] Watson J.D. (1976). *Molecular Biology of the Gene.* 3rd Ed., Menlo Park, CA: Benjamin-Cummings. Cited from *Molecular Biology of the Gene.* (Russia, 1978), Moscow, Mir, p. 281-412.

[49] Tuzikov F.V., Zinoviev V.V., Vavilin V.I., Malygin E.G. (1988). Small-angle X-ray scattering study of enzyme-substrate interaction in solution. *Studia Biophisica.* V. 125, p. 169-172.

[50] Tuzikov F.V., Tuzikova N.A., Vavilin V.I., Zinov'ev V.V., Malygin E.G., Favorova O.O., Zargarova T.A., Sudomoina M.A., Kiselev L.L (1991). Aggregation of tryptophanyl-tRNA synthetase depending on temperature. Small-angle X-ray scattering study. *Molecular Biology* (Russia). V. 25, p. 740-751.

[51] Studitsky V.M., Kassavetis G.A., Geiduschek E.P., Felsenfeld G. (1997). Mechanism of transcription through the nucleosome by eukaryotic RNA polymerase. *Science.* V. 278, p. 1960-1963.

[52] Studitsky V.M. (2001). Transcription through chromatin. *Molecular Biology* (Russia). V. 35, p. 235-247.

[53] Semenov M.A., Gasan A.I., Bolbukh T.V., Maleev V.Ya. (1996). Hydration and structural transitions in DNA from Micrococcus Lysodeikticus in films. *Biophysics* (Russia). V. 41, p. 1007-1016.

[54] Semenov M.A., Maleev V.Ya., Bereznyak E.G., Gasan A.I., Bolbukh T.V. (1991). *Molecular Biology* (Russia). V. 25, p. 1626-1634.

[55] Maleev V.Ya., Semenov M.A., Gasan A.I., Kashpur V.A. (1993). Physical properties of the DNA-water system. *Biophysics* (Russia). V. 38, p. 768-791.

[56] Panin L.E., Kunitsyn V.G., Tuzikov F.V. (2005). Effect of glucocorticoids and their complexes with apolipoprotein A-I on the secondary structure of eukaryotic DNA. *Int. J. Quantum Chem.*. V. 101, p. 450-467.

[57] Starikov E.B., Semenov M.A. (1988). *J. Phys. Chem.* (Russia). V. 42, p. 2120-2126.

[58] Sukhorukhov B.I., Sukhorukhov G.B., Shabarchina L.I., Montrel M.M. (1996). Optical properties and structure of Langmuir films of nucleic acid complexes with lipids and synthetic amphiphilic molecules. 2. IR spectrum, hydration, and conformation state of DNA in the multilayer Langmuir film of its complex with octadecylamine. *Biophysics* (Russia). V. 41, p. 1016-1025.

[59] Vilkov L.V., Pentin Yu.A. (1987). *Physical Methods in Chemistry.* Vyschaya Shkola, 368 p.

[60] Sawyer N.J., Oliver J.T., Troop R.S. (1963). Observation on the role of the RES in the metabolism of adrenocortical steroids. *Steroids.* V. 2, p. 213-227.

[61] Kadesch T.R., Chamberlin M.J. (1982). Studies of in vitro transcription by calf thymus RNA polymerase II using a novel duplex DNA template. *J. Biol. Chem.,* V. 257, p. 5286-5295.

[62] Wingender E. (1997). The classification of eukaryotic transcriptional factors. *Molecular Biology* (Russia). V. 31, p. 584-6000.

[63] Rowles J., Scherer S.W.Xi, Majer T., Nickle M., Rommens D.C., Popov J.M., Harris K.M., Riebow R.A., Xia N.L., Tsui J., Bogardus L.C., Prochazka C. (1996). Cloning and characterization of PDK4 on 7q21.3 encoding a fourth pyruvate dehydrogenase kinase isoenzyme in human. *J. Biol. Chem.* V. 271, N 37, p. 22376-22382.

[64] Perez-Castro A.V., Wilson J., Altherr M.R. (1997). Genomic organization of human fibroblast growth factor receptor 3 (FGFR3) gene and comparative sequence analysis with the mouse Fgfr3 gene. *Genomics*. V. 41, N 1, p. 10-16.

In: Trends in DNA Research
Editor: Corey R. Woods, pp. 103-122
ISBN 1-59454-635-5


*Chapter V*

# Mammalian D-Amino-Acid Oxidase, D-Serine and Serine Racemase: Recent Progress

***Ryuichi Konno***
Department of Microbiology, Dokkyo University School of Medicine, Mibu, Japan

## Abstract

D-amino-acid oxidase (DAO) which catalyzes oxidative deamination of D-amino acids was found nearly 70 years ago in mammals. However, since D-amino acids are considered very rare in mammals, the physiological role of DAO has been enigmatic. Development of analytical methods has revealed that D-amino acids are present in considerable amounts even in mammals. Therefore, it becomes clear that the metabolism of these D-amino acids is one of the functions of DAO. In addition, brain DAO is shown to have a function to modulate neurotransmission. D-Serine in the brain is shown to bind to the N-methyl-D-aspartate (NMDA) subtype glutamate receptors to potentiate the neurotransmission. By catabolizing this D-serine, DAO is considered to further regulate the neurotransmission. The origin of D-serine was not clear, but serine racemase which converts L-serine to D-isomer has been finally discovered in the brains of rodents and humans. Recently, DAO is shown to be implicated in schizophrenia. The newly discovered gene product activates DAO and reduces D-serine, resulting in the hypofunction of NMDA receptors. This hypofunction is likely a cause of schizophrenia. These findings have been resolving the enigma which lasts over 70 years.

# Introduction

D-Amino-acid oxidase (DAO, EC 1.4.3.3) catalyzes oxidative deamination of D-amino acids, stereoisomers of naturally occurring L-amino acids (Fig. 1). In the course of the reaction, the corresponding 2-oxo acids, hydrogen peroxide and ammonia are produced (Equation 1). This enzyme was discovered by Sir Hans A. Krebs around 1935. Since D-amino acids have been considered to be rare in eukaryotes, especially in mammals, the physiological role of DAO has been enigmatic for a long time (see Meister, 1965; see Brosnan, 2001). However, improvements and development of analytical methods have shown that D-amino acids are present in mammals and have some physiological functions (see Imai et al., 1996). Accordingly, the functions of DAO have been gradually revealed.

$$\mathrm{R-\underset{H}{\overset{COOH}{C}}-NH_2 + O_2 + H_2O \longrightarrow R-\overset{COOH}{C}=O + H_2O_2 + NH_3}$$

D-Amino acid

2-Oxo acid

Equation 1
Reaction catalyzed by D-amino-acid oxidase

$$\mathrm{HO-H_2C-\underset{H}{\overset{COOH}{C}}-NH_2 \longleftrightarrow H_2N-\underset{H}{\overset{COOH}{C}}-CH_2-OH}$$

D-Serine

L-Serine

Equation 2
Reaction catalyzed by serine racemase

It may be suspected that DAO does not function *in vivo* like *in vitro*. However, DAO, in deed, works *in vivo* because the lack of this enzyme causes the accumulation of a large amount of D-amino acids in the organs and body fluid of mice (see below). The kidney is the richest organ of DAO in the body and, therefore, the patients with renal disorders have higher content of D-amino acids in the serum than the normal people (Nagata et al., 1987; Brückner and Haush, 1993). The younger animals which have low DAO activity in organs have higher D-amino acid content than the adult animals. D-Amino acids are utilized as nutrients in place of the corresponding L-amino acids in humans and animals where D-amino acids are shown to be converted into 2-oxo acids (see below). All these results indicate that DAO functions in the body.

Recent progress on mammalian DAO, D-serine and serine racemase has been reviewed here. Structure and reaction mechanism of DAO has been reviewed in detail by Pilone (2000). D-Serine detected in mammalian brains is reviewed by Hashimoto and Oka (1997).

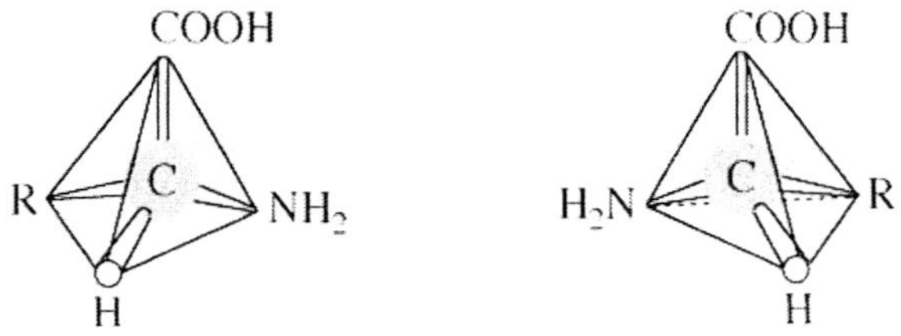


D-Amino acid L-Amino acid

Figure 1. Configuration of D-amino acid and L-amino acid. An amino acid has an asymmetric carbon atom at the center. If four different atoms or groups bind to the carbon atom, two types of stereoisomer are formed as shown in the figure. Physico-chemical properties of D-amino acid and L-amino acid are the same, but organisms easily discriminate them. L-Amino acids are naturally occurring amino acids and D-amino acids have been considered unnatural amino acids.

# Detection of DAO

Presence of DAO can be demonstrated by various methods (Konno, 1998a). Traditionally the presence of DAO has been shown by measuring enzyme activity. Since DAO has a wide spectrum of substrate specificity (Dixon and Kleppe, 1965b), many neutral D-amino acids are used as substrates. D-Alanine, D-proline, D-methionine, and D-phenylalanine are often used. Since 2-oxo acid, hydrogen peroxide, and ammonia are produced in the reaction (Equation 1), DAO activity can be determined by quantitating one of these products. A large number of enzyme activity assays have been reported. Assays which measure the product 2-oxo acid are described by Biondi et al. (1991) and D'Aniello et al. (1993). Assays which quantitate hydrogen peroxide are described by Duley and Holmes (1975) and Watanabe et al. (1978). An assay method which measures ammonia is described by Sasamura et al. (2002). These are some representatives.

DAO activity can be also determined by measuring oxygen consumption using a manometer (Kun et al., 1954) or an oxygen electrode (Dixon and Kleppe, 1965a; Gavazzi et al., 1987). However, these methods are restricted to a specific application nowadays.

The presence of DAO is also detected histochemically and cytochemically using the enzyme activity (Arnold et al., 1979; Horiike et al., 1987, 1994). Immunohistochemical method using antibody against DAO is used to localize DAO in the tissues, cells and organella (Usuda et al., 1986, 1991; Perotti et al., 1987; Johkura et al., 1998; Moreno et al., 1999).

Using antibody against DAO, it is also possible to measure the amount of DAO molecules by enzyme-linked immunosorbent assay (ELISA) (Kawasaka and Tatsumi, 1998). Immunoblotting or western blotting using antibody against DAO has shown the presence of DAO in various organisms (Gavassi et al., 1987) and in various organs (Konno et al., 1991b, 1997).

Since nucleotide sequences of DAO cDNA and the DAO gene are known, it is possible to detect DAO mRNA and to determine the expression level of the DAO gene under various conditions. Northern hybridization has been mainly used to examine the size and the expression level of DAO mRNA (Fukui et al., 1987; Momoi et al., 1988; Tada et al., 1990; Konno et al., 1997; Konno, 1998b). Reverse transcription-PCR (RT-PCR) can be used to examine the presence of the message for DAO (Konno et al., 1997) and further to determine the relative abundance of the message (Sarower et al., 2003). It is possible to use real-time PCR for the quantitation of DAO mRNA. However, this method is not used yet. Using *in situ* hybridization, it is possible to localize DAO mRNA at the tissue level (Koibuchi et al., 1995).

## Localization of DAO

DAO is present in a wide range of organisms from microorganisms to humans (see Meister, 1965). In mammals, DAO is mainly localized in the kidney, liver and brain. It is present in epithelial cells of proximal tubules in the kidney (Angermüller, 1989) and hepatocytes of the liver (Perotti et al., 1987; Usuda et al., 1991). However, it should be emphasized that the mouse does not have DAO in the liver but the reason is not known (Konno et al., 1997). In the brain, DAO is localized in the midbrain, pons, medulla oblongata, cerebellum, and spinal cord (Horiike et al., 1994). DAO is present in astrocytes but not in neurons (Arnold et al., 1979; Horiike et al., 1987, 1994). In the cerebellum, DAO is also present in Bergmann glial cells (Horiike et al., 1987). However, Moreno et al. (1999) indicated that DAO was present in wider area of the brain and was also present in neurons. Since this is a very important point, it has to be precisely determined whether or not neurons have DAO.

Intracellularly DAO is localized in the organelle, peroxisomes, together with catalase, other hydrogen peroxide producing enzymes and enzymes involved in the fatty acid β-oxidation (Angermüller, 1989). It is because DAO molecules have the peroxisome-targeting signal (Ser-His-Leu) at the carboxy terminus (Gould et al., 1988).

## Properties of DAO

DAO is a single polypeptide containing flavin adenine dinucleotide (FAD) as a prosthetic group. Therefore, DAO has a tinge of yellow. FAD is not covalently bound to the peptide and is easily released during the purification of mammalian DAO.

Mammalian DAO is not an inducible enzyme. Even when a large amount of D-amino acids were loaded, DAO activity in the kidney and liver was not elevated (Nagata et al., 1991). However, an exception was observed in very young animals. D'Aniello et al. (1993) found that when a mother rat ingested D-alanine during pregnancy and suckling, DAO activity in the liver and kidneys of the baby rats was increased. In contrast to mammalian DAO, fish and yeast DAOs are easily induced by the environmental D-amino acids (Sarower et al., 2003; Molla et al., 2003).

DAO has a wide spectrum of substrate specificity and metabolizes many neutral D-amino acids. However, acidic and basic D-amino acids are not catabolized. Stereospecificity is so strict that L-amino acids are not metabolized. The reason why L-amino acids are not used as a substrate is explained by the structure of the active site of the DAO molecule and the reaction mechanism. The three-dimensional structures of pig and yeast DAO have been clarified and the catalytically important amino acid residues are identified (Mattevi et al. 1996; Mizutani et al. 1996; Tadone et al., 1997; Setoyama et al., 2002; Pollegioni et al., 2002).

## Molecular Biology

Human (Momoi et al., 1988), pig (Ronchi et al., 1982; Fukui et al., 1987), rabbit (Momoi et al., 1990), and guinea pig (Konno et al., 1999) DAO consist of 347 amino acid residues. However, rat (Konno, 1998b) and hamster (Tsuchiya et al., 2003) DAO consist of 346 amino acid residues whereas mouse DAO consists of 345 amino acid residues (Tada et al., 1990). Interestingly carp DAO also has 347 amino acid residues (Sarower et al., 2003). Therefore, during the evolution of mammals, it seems that the one amino acid residue was lost in the rodent lineage and that another residue was further lost in the lineage to the living mouse. However, mammalian DAOs are highly homologous and more than 80% of amino acid residues are identical. The important residues have been conserved through the evolution. The nucleotide sequence of DAO cDNA and amino acid sequences of DAO are deposited in GenBank/EMBL/DDBJ under the following accession numbers: X13227 (human), M16972 (pig), D12494 (rabbit), AJ007710 (guinea pig), AB00340 (rat), AB016530 (hamster), M32299 (mouse), AY421707 (carp), and U60066 (yeast *Rhodotorula gracilis*).

The DAO gene is present in a single copy in the human, rat and mouse genome. All of these DAO genes span approximately 20 kb and consist of 11 exons divided by 10 introns. The gene organization is very similar among these mammals. The DAO gene is mapped at human chromosome 12q24, rat chromosome 12q16 and mouse chromosome 5q3-F (Fukui and Miyake, 1992; Konno, 2001). These chromosomal regions are syntenic among these mammals. From the analysis of recombinant inbred strains and linkage analysis, the mouse DAO locus is mapped at 65 cM on the chromosome 5 (Hilgers and Arends, 1985; Konno et al, 1989b). Restriction fragment length polymorphism (RFLP) in the mouse DAO gene was examined using eight restriction enzymes and 15 inbred strains (Konno et al., 1996).

## DAO and Detoxication

It is natural to consider that DAO is a detoxication enzyme because the presence of antipoles in the homochiral world would disturb the harmony. However, DAO would not be a detoxifying enzyme because most D-amino acids are non-toxic (see Harper et al., 1970; Schieber et al., 1997) and D- or L-configuration of amino acids is strictly discriminated in the body. D-Amino acids ingested are either metabolized or excreted into urine (Konno et al., 1984, 1988b; Schieber et al., 1997). Lehmann et al. (1983) showed that one third of orally administered D-phenylalanine was converted into L-phenylalanine while one third was excreted into urine in humans. The facts that mutant ddY/DAO$^-$ mice lacking DAO do not manifest significant pathological symptoms and live long as do control ddY/DAO$^+$ mice seem to support the idea that D-amino acids are not toxic and DAO is not a detoxifying enzyme.

D'Aniello et al. (1993) suggested that a large amount of D-amino acids might cause retardation of growth. The growth retardation, however, would not be due to the toxic effects of D-amino acids but due to the effects of disproportionate amounts of amino acids. It is known that L-amino acids in excess exert the same or more adverse effects than D-amino acids (see Harper et al., 1970).

Most D-amino acids are non-toxic but one clear exception is D-serine. It is well known that D-serine is nephrotoxic in rats (Ganote et al., 1974; Carone and Ganote, 1975). D-Serine causes acute necrosis of cells in the proximal tubules, resulting in polyuria, glucosuria, aminoaciduria, and proteinuria. It is not known whether DAO is involved in this nephrotoxicity. DAO may not be involved in this nephrotoxicity because this phenomenon occurs only in rats. D-Serine is not nephrotoxic in mice, hamsters, guinea pigs, gerbils, rabbits or dogs (Morehead et al., 1946; Kaltenbach et al., 1979). If hydrogen peroxide produced by the DAO reaction is the cause of the injury (Silbernagl et al., 1999), the same phenomenon should have been observed in other animals as well. Mutant rats lacking DAO may be able to resolve this problem.

On the contrary, DAO becomes a cause of nephrotoxicity in some case. D-Propargylglycine is nephrotoxic in rats and mice (Nakajima et al., 1981; Konno et al., 2000). However, it was found that D-propargylglycine itself was not toxic but became toxic when it was metabolized by DAO. D-Propargylglycine caused necrosis of the cells in the proximal tubule in normal ddY/DAO$^+$ mice and brought polyuria, glucosuria, aminoaciduria, and proteinuria, whereas none of these symptoms was observed in mutant ddY/DAO$^-$ mice lacking DAO (Konno et al., 2000).

## Mutant Mice Lacking DAO

When we were examining mice of various strains in search of mutants having disorders in DAO, we found some mice in the population of non-inbred ddY mice lacked DAO activity (Konno and Yasumura, 1983). A simple and efficient method to screen for these mice was devised (Konno and Yasumura, 1988). A mutant strain lacking DAO activity (ddY/DAO$^-$) together with a control strain (ddY/DAO$^+$) which possesses DAO activity have been established. The ddY/DAO$^-$ mice carry a missense mutation in the DAO gene which causes a substitution of an amino acid residue (Gly181Arg) (Sasaki et al., 1992). These mice had a DAO protein which did not have catalytic activity (Konno et al., 1991b). Inactive DAO was found in the peroxisomes because the DAO molecule had an intact targeting signal at the carboxyl terminus (Konno et al., 1991b; Sasaki et al., 1992).

Significant amounts of D-amino acids were detected in the organs, serum and urine of the mutant ddY/DAO$^-$ mice. First, D-methionine was found in the urine (Konno et al., 1988b). Then, D-alanine was detected in the urine, cerebrum, cerebellum, medulla oblongata, hippocampus, pituitary gland, hypothalamus, pineal gland, and serum (Konno et al., 1989a: Nagata, 1992; Hashimoto et al., 1993b; Nagata et al., 1994; Morikawa et al., 2001). D-Serine was also found in the cerebrum, cerebellum, kidney, serum, and urine (Konno et al., 1991a; Nagata, 1992; Hashimoto et al., 1993b; Nagata et al., 1994). D-Proline was present in the kidney, cerebellum, pituitary gland, serum, and urine (Konno et al., 1991a; Nagata, 1992;

Nagata et al., 1994; Hamase et al., 2001). D-Leucine was detected in the cerebrum, cerebellum, medulla oblongata, hippocampus, pituitary gland, hypothalamus, pineal gland, and serum (Hamase et al., 2001). Some of these D-amino acids were also found in the normal ddY/DAO$^{+}$ mice possessing DAO but the amounts were far less than those in the ddY/DAO$^{-}$ mice. D-Aspartic acid was also detected in the kidney, cerebrum, cerebellum, medulla oblongata, hippocampus, pituitary gland, hypothalamus, pineal gland, and serum of ddY/DAO$^{-}$ mice (Nagasaki et al., 1990; Hashimoto et al., 1993b). However, the content of D-aspartic acid was not different between the ddY/DAO$^{-}$ and ddY/DAO$^{+}$ mice. This is reasonable because DAO attacks only neutral D-amino acids and does not catabolize acidic D-aspartic acid. D-Aspartic acid is metabolized by D-aspartate oxidase (E.C. 1.4.3.1), a sister enzyme of DAO having the same evolutional origin (Negri et al., 1992; Simonic et al., 1997; Setoyama and Miura, 1997). It was shown that the ddY/DAO$^{-}$ mice had the same level of D-aspartate oxidase activity as ddY/DAO$^{+}$ mice (Nagasaki et al., 1990).

## D-Aminoaciduria

The ddY/DAO$^{-}$ mice excreted a large amount of methionine into urine (Konno et al., 1988b). Then, methionine was purified from the urine and was applied to high-performance liquid chromatography (HPLC) to separate D- and L-isomer. This was the first case that HPLC was used to split D- and L- isomer of amino acids derived from biological sample. The analysis showed that most of the urinary methionine was D-methionine (Konno et al., 1988b). The result was astonishing because it was believed at that time that D-amino acids were not present in eukaryotes, especially in mammals. Therefore, the source of the D-methionine was pursued. Soon our analysis showed that the mouse diet contained D-methionine. It became clear that the commercial diets were supplemented with DL-methionine to meet the nutritional requirement of experimental animals. Since the ddY/DAO$^{-}$ mice were unable to metabolize the absorbed D-methionine due to the lack of DAO, they excreted it into urine. Therefore, when they were given a special diet in which supplementary DL-methionine was replaced by L-methionine, the urinary D-methionine became undetectable (Konno et al., 1988b).

The ddY/DAO$^{-}$ mice also excreted a large amount of D-alanine into urine. However, it was not a dietary origin (Konno et al., 1989a). When the ddY/DAO$^{-}$ mice were given antibiotics, the excretion of D-alanine became very low (1990). This result suggested the bacterial origin of the urinary D-alanine since bacterial cell wall (peptidoglycan) contains D-alanine as an essential constituent. To confirm this, the ddY/DAO$^{-}$ mice were made germ-free. The germ-free ddY/DAO$^{-}$ mice scarcely excreted D-alanine. However, when they were made gnotobiotic by inoculation of two species of intestinal bacteria, they started to excrete a large amount of D-alanine into urine again. The gnotobiotic mice further inoculated with two additional species of intestinal bacteria excreted more D-alanine into urine (Konno et al., 1993). These results clearly indicated that the origin of the urinary D-alanine was intestinal bacteria.

Urine of the ddY/DAO$^{-}$ mice also contained D-serine. It was found that most of the D-serine came from their diet but some part was of internal origin (Asakura and Konno, 1997). However, its source was not clear until the discovery of serine racemase (Wolosker et al.,

1999a). It is considered that D-serine is synthesized by serine racemase in the bodies of the ddY/DAO$^-$ mice.

## Nutritional Utilization of D-Amino Acids

Some D-amino acids are utilized in place of the essential L-amino acids (see Meister, 1965). These D-amino acids are at first converted to the corresponding 2-oxo acids by DAO. Then, the 2-oxo acids are asymmetrically reaminated to L-amino acids by ubiquitous transaminase. This two-step conversion was confirmed by the nutritional experiments employing the ddY/DAO$^-$ mice. In contrast to the ddY/DAO$^+$ mice, the ddY/DAO$^-$ mice could not utilize D-phenylalanine in place of L-phenylalanine, though these two strains of mice could utilize phenylpyruvate (2-oxo acid derived from phenylalanine) in place of L-phenylalanine (Konno and Yasumura, 1984). These results indicate that DAO is an indispensable enzyme for the utilization of D-amino acids and that no other metabolic pathways which are responsible for the utilization of D-amino acids exist in mice.

The above conclusion obtained the support from the experiments using the cell culture. Rat liver cells cultivated for a long period of time *in vitro* could not grow in the culture medium in which essential L-phenylalanine was replaced with D-phenylalanine. However, when DAO was induced, these cells became able to grow in the medium containing D-phenylalanine and they had a plenty of L-phenylalanine in their cellular proteins (Yoda et al., 2001).

Using D-leucine labeled with stable isotope, Hasegawa et al. (2004) more directly examined the conversion of D-leucine to α-keto-valeric acid (2-oxo acid derived from leucine) and further to L-leucine. This conversion occurred in the ddY/DAO$^+$ mice but not in the ddY/DAO$^-$ mice. They also showed that the conversion was significantly slowed down in partially nephrectomized rats. These results indicated that DAO is an indispensable enzyme for the conversion of D-leucine to 2-oxo acid and the kidney is the principal organ for this conversion.

The ddY/DAO$^-$ mice accumulated D-methionine of dietary origin in their organs and finally excreted it into urine. This result indicates that the D-methionine is constantly metabolized by DAO in normal mice. Therefore, one of the physiological functions of DAO is the metabolism of dietary D-amino acids. It is well known that foods produced by microbial fermentation contain a large amount of various D-amino acids (Brückner et al., 1995). Even natural products such as vegetables and fruits contain D-amino acids (Brückner and Westhauser, 1994). Alkaline treatment of foods and heat during the cooking cause racemization of amino acids (see Mann and Bada, 1987). Therefore, we daily ingest a good amount of foods containing D-amino acids. These D-amino acids are absorbed in the intestine and are transported to the liver and kidney through bloodstream. Some of them are metabolized by DAO and the resultant 2-oxo acids are utilized in the metabolism.

# D-Serine in the Brain

Hashimoto et al. (1992) found that substantial quantity of D-serine was present in the rat brain. It was a surprising finding. Nagata (1992) also showed the presence of D-serine in the mouse brain. She showed the D-serine content was high in the cerebrum and was low in the cerebellum. Since the D-serine content in the cerebellum was as high as in the cerebrum in the ddY/DAO$^-$ mice, she concluded that this distribution pattern was created by DAO. Subsequently, the presence of D-serine was confined to narrow areas of the brain. Hashimoto et al. (1993c) found that D-serine was present in the forebrain area of rodents, such as cerebral cortex, hippocampus and striatum. Nagata et al. (1994) also showed that D-serine was rich in the frontal lobe, occipital lobe, hippocampus, hypothalamus, and olfactory bulb in the rat brain. This phenomenon was not restricted to the rodents. Human brains also contained high level of D-serine (Chouinard et al., 1993; Hashimoto et al., 1993a).

Hashimoto et al. (1993c) noticed that the distribution of D-serine was parallel with that of *N*-methyl-D-aspartate (NMDA) subtype of glutamate receptors on the surface of neurons. They suggested that D-serine was a ligand for the strychinine-insensitive "glycine" site of the NMDA receptors (this site was called glycine site because only glycine was thought to bind to this site).

Using antibody against D-serine, Schell et al. (1995) clearly demonstrated the distribution of D-serine in the rat brain. Highest levels of D-serine were observed in the cerebral cortex, hippocampus, anterior olfactory nucleus, olfactory tubercle and amygdale. This distribution pattern was similar to that of NMDA receptors and inversely correlated with that of DAO. D-Serine was found to be present in protoplasmic astrocytes in grey matter. In glial cultures of rat cerebral cortex, D-serine was rich in type 2 astrocytes and was released from these cells when they were stimulated by agonists of non-NMDA glutamate receptors. Later, Schell et al. (1997) showed that the distribution pattern of D-serine closely resembled that of NR2A/B subtypes of NMDA receptors.

Yasuda et al. (2001) also examined the presence of D-serine using antibody against D-serine. They found that D-serine was present not only in astrocytes but also in some neurons. This result was different from that of Schell et al. (1995). Since this is an important matter, it should be carefully examined whether neurons have D-serine.

It is known that for the full activation of NMDA receptor, in addition to the binding of glutamate, simultaneous binding of co-agonist (glycine or D-serine) to the glycine site of the receptor is required. Using cloned NMDA receptors expressed on the surface of oocytes, Matsui et al. (1995) found that D-serine was more potent than glycine in activation of these NMDA receptors.

Administration of NMDA (co-agonist of glutamate) to the cerebellum of freely-moving rats via a microdialysis probe increased extracellular cGMP level through activation of the NMDA receptor/nitric oxide pathway. D-Serine potentiated this NMDA-evoked cGMP production in a concentration-dependent manner (Fedele et al., 1997). These results are consistent with the idea that full activation of the NMDA receptors requires co-activation of the glutamate site and of the glycine site.

Using antibody against D-serine, Wako et al. (1995) observed that D-serine injected into the lateral ventricle accumulated in glial cells in the rat brain, though they could not detect D-serine-containing cells in the brains of control rats. Yamamoto et al. (2001) also showed D-serine was taken up by synaptosomal P2 fraction isolated from rat brain. These results suggest the presence of a transporter system for D-serine.

From the compiled data, Snyder and his colleagues have proposed a hypothesis that astrocyte-derived D-serine binds to the glycine site of the NMDA receptor and modulates the neurotransmission (see Snyder and Kim, 2000).

Tsuchida et al. (2001) administered D-serine into infant rats and searched for transcripts whose levels were up-regulated using RNA fingerprinting. They cloned a novel transcript, *dsr-1* (D-serine responsive transcript-1). Since the deduced amino acid sequence had a membrane spanning domain, they suggested the *dsr-1* product involved in the regulation of vesicular or synaptic functions. The precise nature and function of this molecule is expected to be determined.

## Serine Racemase

For a long time, the origin of D-serine present in the brain was obscure though *de novo* synthesis was speculated. Wood et al. (1996) found that D-serine was formed from L-phosphoserine in brain synaptosomes but the precise mechanism was not determined. Iwama et al. (1997) showed a glycine cleavage system was implicated in biosynthesis of D-serine but this possibility was not pursued further. Then, Dunlop and Neidle (1997) found that radiolabeled L-serine was converted to D-serine in the rat brain and suggested the presence of racemase which directly converts L-serine to D-serine (Equation 2).

Uo et al. (1998) partially purified serine racemase from silk worms for the first time in eukaryotes. They found that pupa homogenate did not show racemase activity but ammonium sulfate fractionation of the homogenate revealed racemase activity. Using the same strategy, Wolosker et al. (1999a) purified serine racemase from rat brains. It had pyridoxal 5'-phosphate as a prosthetic group like serine racemase from the silk worm. It was highly selective toward L-serine and did not racemize other L-amino acids. Wolosker et al. (1999b) then cloned mouse serine racemase. Mouse serine racemase was found to consist of 339 amino acid residues and have a molecular mass of 36,359. Since robust synthesis of D-serine was occurred in cultured cells transfected with serine racemase cDNA, it is certain that the cloned cDNA encoded for functional serine racemase. Serine racemase was co-localized with D-serine in protoplasmic astrocytes, indicating that D-serine is synthesized by serine racemase within these cells. However, one strange thing was observed in the northern blot analysis and western blot analysis. Northern blot showed a single message of 2.6 kb in the liver, brain and kidney. The message level was extremely higher in the liver than in the brain. However, on the contrary, western blot showed serine racemase was more abundant in the brain than in the liver. It would be important to solve this discrepancy.

De Miranda et al. (2000) cloned human serine racemase. Human serine racemase had a molecular mass of 36,566 consisting of 340 amino acid residues. Serine racemase was expressed in hippocampus, corpus callosum, substantia nigra and caudate. The serine racemase gene (*SRR*) consisted of 8 exons and was localized at chromosome 17q13.3 (De Miranda et al., 2000). Mouse serine racemase gene (*Srr*) also consisted of 8 exons and was localized on chromosome 11. Rat serine racemase was recently cloned (Konno, 2003). Rat serine racemase had a molecular mass of 35,691 consisting of 333 amino acids residues. It was slightly smaller than the mouse and human enzymes because it was truncated at the carboxyl terminus. It is interesting that both rat and mouse enzymes have a common deletion of the 322nd amino acid residue compared with human serine racemase. It is a quite coincidence that rat and mouse DAOs also have a common deletion of an amino acid residue compared with other mammalian DAOs. However, serine racemase of these mammals are highly homologous and 92% (human *vs.* rat), 88% (human *vs.* mouse), and 96% (rat *vs.* mouse) of amino acid residues are identical.

The nucleotide sequence of serine racemase cDNA and amino acid sequences of serine racemase are deposited in GenBank/EMBL/DDBJ under the following accession numbers: AF169974 (human), AB106282 (rat), and AF148321 (mouse).

Serine racemase functions as a homodimer (Neidle and Dunlop, 2002; Stříšovský et al., 2003). This enzyme was shown to convert L-serine not only to D-serine but also to pyruvate (Neidle and Dunlop, 2002; De Miranda et al., 2002; Stříšovský et al., 2003). This may explain why the serine racemase gene are ubiquitously expressed (Wolosker et al., 1999b; Konno, 2003) though D-serine is exclusively present in the brain. However, the synthesis of D-serine by serine racemase is not firmly established. There is a possibility that another mechanism to produce D-serine may be present in the brains.

## Brain DAO

Using brain slices and hippocampal cell cultures, Mothet et al. (2000) showed that exogenously applied DAO attenuated NMDA receptor-mediated synaptic currents by the selective degradation of D-serine and that this inhibition was reversed by the exogenously applied D-serine.

The mutant ddY/DAO$^-$ mice showed a strong response to the chronic nociceptive stimuli (Wake et al., 2001). These phenomena can be explained by the high concentration of D-serine in the mutant mice due to the lack of DAO, resulting in the increase of the binding to NMDA receptors and the enhanced activation of the receptors. In deed, ddY/DAO$^-$ mice exhibited an augmented excitatory post-synaptic currents mediated by NMDA receptors in spinal cord dorsal horn neurons (Wake et al., 2001).

Stevens et al. (2003) detected D-serine and serine racemase in astrocytes and Müller cells in the retina of rats, mice and salamanders. In ganglion cells isolated from rat retina, D-serine enhanced excitatory currents elicited by application of NMDA. D-Serine also enhanced NMDA component of light-evoked synaptic responses in the salamander retina. Exogenously applied DAO reduced the magnitude of the NMDA receptor-mediated currents.

All these results are in agreement with the hypothesis that D-serine functions as an endogenous coagonist at the glycine site of NMDA receptors and modulates the neurotransmission. In addition, DAO further modulates the neurotransmission by controlling the concentration of the D-serine (Fig. 2). Stimulation and inhibition are the mechanism to keep the system in our body functioning properly.

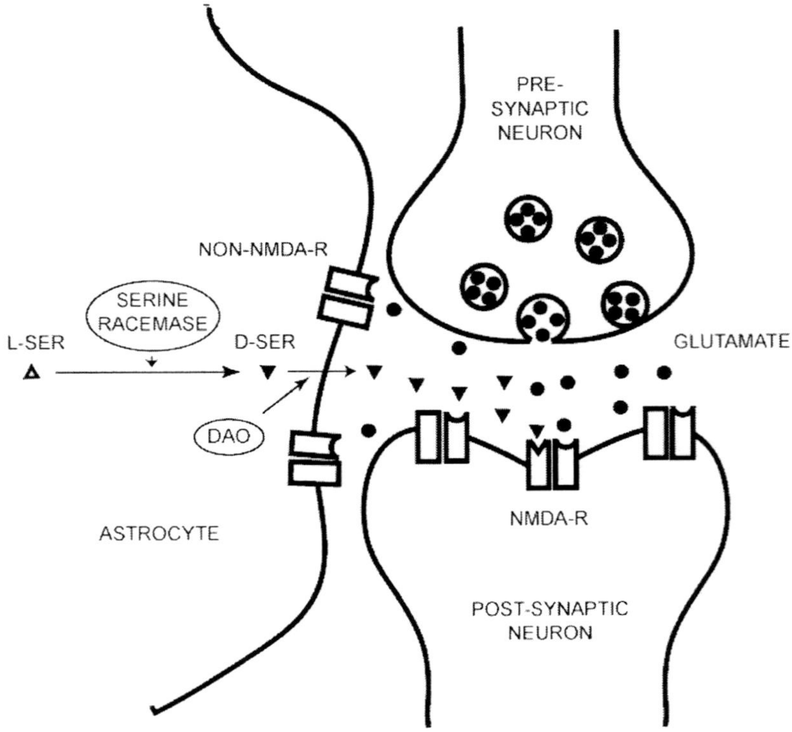


Figure 2. Model showing the roles of D-serine, serine racemase and D-amino-acid oxidase (DAO) in glutamatergic neurotransmission. Glutamate (closed circle) released from pre-synaptic neurons binds to the NMDA receptor (NMDA-R) of post-synaptic neurons and to the non-NMDA receptors (non-NMDA-R) on astrocytes. The activated astrocytes release D-serine (closed triangle). Binding of D-serine together with glutamate to the NMDA receptors activates the post-synaptic neuron. The NMDA receptors require the binding of two ligands for the full activation. D-Serine is converted from L-serine (open triangle) by serine racemase in astrocytes. DAO metabolizes D-serine and reduces the concentration of D-serine, resulting in the attenuation of the activation of NMDA receptors. This model is based on that of Snyder and Kim (2000).

## Schizophrenia

Recently an astonishing article has been published concerning the function of DAO. Chumakov et al. (2002) constructed a map of single-nucleotide polymorphisms (SNPs) associated with schizophrenia and found a novel gene, *G72* at chromosome 13q34 which encoded a 153-amino acid protein, was implicated in this mental disease. Yeast two-hybrid experiments identified DAO as an interacting partner of the *G72* product. DAO itself had a week association with schizophrenia but the combination of *G72* and *DAO* was found to have a synergistic effect on this disease. They confirmed that the *G72* product activated DAO *in vitro*. They consider that the *G72* product activates DAO and decreases the level of D-serine, resulting in hypofunction of the NMDA receptors. Hypofunction of NMDA receptors is recognized as one of the most plausible causes of schizophrenia.

This hypothesis seems probable since schizophrenic patients who received D-serine together with antipsychotics showed significant improvements in their positive, negative, and cognitive symptoms (Tsai et al., 1998). The results that the D-serine content in the serum was lower in patients with schizophrenia than in healthy controls (Hashimoto et al., 2003) also agree with this hypothesis.

## Concluding Remarks and Perspective

Physiological roles of DAO have been unknown for a long time. However, at least some of them have been unveiled. Metabolism of D-amino acids of internal and external origin is one of the functions of hepatic and renal DAO. To some extent, D-amino acids are utilized as nutrients. On the contrary, brain DAO is involved in regulation of the neurotransmission by degrading D-serine which binds to the NMDA receptors. Serine racemase which converts L-serine to D-serine has been finally discovered. DAO, D-serine and serine racemase have opened a new field in the neuroscience. Recently, DAO is shown to be implicated in schizophrenia. It is considered that the product of the newly discovered *G72* gene activates DAO, which decreases the D-serine level. The low level of D-serine causes the hypofunction of NMDA receptors. This hypofunction of the NMDA receptors is likely the cause of the schizophrenia. However, this hypothesis needs to be rigorously verified. It also needs careful examination whether D-serine is exclusively synthesized by serine racemase.

Since NMDA receptors are involved in many aspects of brain physiology and pathology, DAO may be involved in more brain functions. In addition, since the kidney and liver have DAO more than enough to metabolize D-amino acids of external and internal origins, DAO may have other functions yet undiscovered. Mutant ddY/DAO$^-$ mice lacking DAO would be useful for the elucidation of these problems.

## Acknowledgements

Our researches cited in this review were supported in part by grants from the Seki Minato Foundation and the Asahi Glass Foundation.

## References

Angermüller, S. (1989). Peroxisomal oxidases: cytochemical localization and biological relevance. *Prog Histochem Cytochem, 20*, 1-63.

Arnold, G., Liscum, L. & Holtzman, E. (1979). Ultrastructural localization of D-amino acid oxidase in microperoxisomes of the rat nervous system. *J Histochem Cytochem, 27*, 735-745.

Asakura, S. & Konno, R. (1997). Origin of D-serine present in urine of mutant mice lacking D-amino-acid oxidase activity. *Amino Acids, 12*, 213-223.

Biondi, P. A., Guidotti, L., Negri, A. & Secchi, C. (1991). High-performance liquid chromatographic determination of D-amino acid oxidase activity. *J Chromatogr, 566*, 377-382.

Brosnan, J. T. (2001). Amino acids, then and now - a reflection on Sir Hans Krebs' contribution to nitrogen metabolism. *IUBMB Life, 52*, 265-270.

Brückner, H. & Westhauser, T. (1994). Chromatographic determination of D-amino acids as native constituents of vegetables and fruits. *Chromatographia, 39*, 419-426.

Brückner, H., Langer, M., Lüpke, M., Westhauser, T. & Godel, H. (1995). Liquid chromatographic determination of amino acid enantiomers by derivatization with *o*-phtaldialdehyde and chiral thiols. Applications with reference to food science. *J Chromatogr A, 697*, 229-245.

Carone, F. A. & Ganote, C. E. (1975). D-Serine nephrotoxicity. *Arch Pathol, 99*, 658-662.

Chouinard, M. L., Gaitan, D. & Wood, P. L. (1993). Presence of the *N*-methyl- D-aspartate associated glycine receptor agonist, D-serine, in human temporal cortex: comparison of normal, Parkinson, and Alzheimer tissues. *J Neurochem, 61*, 1561-1564.

Chumakov, I., Blumenfeld, M., Guerassimenko, O. et al. (2002). Genetic and physiological data implicating the new human gene G72 and the gene for D-amino acid oxidase in schizophrenia. *Proc Natl Acad Sci USA, 99*, 13675-13680.

D'Aniello, A., D'Onofrio, G., Pischetola, M., D'aniello, G., Vetere, A., Petrucelli, L. & Fisher, G. H. (1993). Biological role of D-amino acid oxidase and D-aspartate oxidase. Effects of D-amino acids. *J Biol Chem, 268*, 26941-26949.

De Miranda, J., Panizzutti, R., Foltyn, V. N. & Wolosker, H. (2002). Cofactors of serine racemase that physiologically stimulate the synthesis of the *N*-methyl-D-aspartate (NMDA) receptor coagonist D-serine. *Proc Natl Acad Sci USA, 99*, 14542-14547.

De Miranda, J., Santoro, A., Engelender, S. & Wolosker, H. (2000). Human serine racemase: molecular cloning, genomic organization and functional analysis. *Gene, 256*, 183-188.

Dixon, M. & Kleppe, K. (1965a). D-Amino acid oxidase. I. Dissociation and recombination of the holoenzyme. *Biochim Biophys Acta, 96*, 357-367.

Dixon, M. & Kleppe, K. (1965b). D-Amino acid oxidase. II. Specificity, competitive inhibition and reaction sequence. *Biochim Biophys Acta, 96*, 368-382.

Duley, J. & Holmes, R. S. (1975). A spectrophotometric procedure for determining the activity of various rat tissue oxidases. *Anal Biochem, 69*, 164-169.

Dunlop, D. S. & Neidle, A. (1997). The origin and turnover of D-serine in brain. *Biochem Biophys Res Comm, 235*, 26-30.

Fedele, E., Bisaglia, M. & Raiteri, M. (1997). D-Serine modulates the NMDA receptor/nitric oxide/cGMP pathway in the rat cerebellum during in vivo microdialysis. *Naunyn-Schmiedeberg's Arch Pharmacol, 355*, 43-47.

Fukui, K. & Miyake, Y. (1992). Molecular cloning and chromosomal localization of a human gene encoding D-amino-acid oxidase. *J Biol Chem, 267*, 18631-18638.

Fukui, K., Watanabe, F., Shibata, T. & Miyake, Y. (1987). Molecular cloning and sequence analysis of cDNAs encoding porcine kidney D-amino acid oxidase. *Biochemistry, 26*, 3612-3618.

Ganote, C. E., Peterson, D. R. & Carone, F. A. (1974). The nature of D-serine-induced nephrotoxicity. *Am J Pathol, 77*, 269-282.

Gavazzi, E., Malgaretti, N. & Curti, B. (1987). Immunochemical properties of D-amino-acid oxidase. *Biochim Biophys Acta, 915*, 188-198.

Gould, S. J., Keller, G.-A. & Subramani, S. (1988). Identification of peroxisomal targeting signals located at the carboxy terminus of four peroxisomal proteins. *J Cell Biol, 107*, 897-905.

Hamase, K., Inoue, T., Morikawa, A., Konno, R. & Zaitsu, K. (2001). Determination of free D-proline and D-leucine in the brains of mutant mice lacking D-amino acid oxidase activity. *Anal Biochem, 298*, 253-258.

Harper, A. E., Benevenga, N. J. & Wohlhueter, R. M. (1970). Effects of ingestion of disproportionate amounts of amino acids. *Physiol Rev, 50*, 428-558.

Hasegawa, H., Matsukawa, T., Shinohara, Y., Konno, R. & Hashimoto, T. (2004). Role of renal D-amino-acid oxidase in pharmacokinetics of D-leucine. *Am J Physiol Endocrinol Metab, 287*, E160-E165.

Hashimoto, A. & Oka, T. (1997). Free D-aspartate and D-serine in the mammalian brain and periphery. *Prog Neurobiol, 52*, 325-353.

Hashimoto, A., Nishikawa, T., Hayashi, T., Fujii, N., Harada, K., Oka, T. & Takahashi, K. (1992). The presence of free D-serine in rat brain. *FEBS Lett, 296*, 33-36.

Hashimoto, A., Kumashiro, S., Nishikawa, T., Oka, T., Takahashi, K., Mito, T., Takashima, S., Doi, N., Mizutani, Y., Yamazaki, T., Kaneko, T. & Ootomo, E. (1993a). Embryonic development and postnatal changes in free D-aspartate and D-serine in the human prefrontal cortex. *J Neurochem, 61*, 348-351.

Hashimoto, A., Nishikawa, T., Konno, R., Niwa, A., Yasumura, Y., Oka, T. & Takahashi, K. (1993b). Free D-serine, D-aspartate and D-alanine in central nervous system and serum in mutant mice lacking D-amino acid oxidase. *Neurosci Lett, 152*, 33-36.

Hashimoto, T., Nishikawa, T., Oka, T. & Takahashi, K. (1993c). Endogenous D-serine in rat brain: *N*-methyl-D-aspartate receptor-related distribution and aging. *J Neurochem, 60*, 783-786.

Hashimoto, K., Fukushima, T., Shimizu, E., Komatsu, N., Watanabe, H., Shinoda, N., Nakazato, M., Kumakiri, C., Okada, S., Hasegawa, H., Imai, K. & Iyo, M. (2003). Decreased serum level of D-serine in patients with schizophrenia. *Arch Gen Psychiatry, 60*, 572-576.

Hilgers, J. & Arends, J. (1985). A series of recombinant inbred strains between the BALB/cHeA and STS/A mouse strains. *Curr Top Microbiol Immunol, 122*, 31-37.

Horiike, K., Tojo, H., Arai, R., Yamano, T., Nozaki, M. & Maeda, T. (1987). Localization of D-amino acid oxidase in Bergmann glial cells and astrocytes of rat cerebellum. *Brain Res Bull, 19*, 587-596.

Horiike, K., Tojo, H., Arai, R., Nozaki, M. & Maeda, T. (1994). D-Amino-acid oxidase is confined to the lower brain stem and cerebellum in rat brain: regional differentiation of astrocytes. *Brain Res, 652*, 297-303.

Imai, K., Fukushima, T., Santa, T., Homma, H., Hamase, K., Sakai, K. & Kato, M. (1996). Analytical chemistry and biochemistry of D-amino acids. *Biomed Chromatogr, 10*, 303-312.

Iwama, H., Takahashi, K., Kure, S., Hayashi, F., Narisawa, K., Tada, K., Mizoguchi, M., Takashima, S., Tomita, U. & Nishikawa, T. (1997). Depletion of cerebral D-serine in non-ketotic hyperglycinemia: possible involvement of glycine cleavage system in control of endogenous D-serine. *Biochem Biophys Res Comm, 231*, 793-796.

Johkura, K., Usuda, N., Liang, Y. & Nakazawa, A. (1998). Immunohistochemical localization of peroxisomal enzymes in developing rat kidney tissues. *J Histochem Cytochem, 46*, 1161-1173.

Kaltenbach, J. P., Ganote, C. E. & Carone, F. A. (1979). Renal tubular necrosis induced by compounds structurally related to D-serine. *Exp Mol Pathol, 30*, 209-214.

Kawasaka, K. & Tatsumi, N. (1998). D-Amino acid oxidase activity in urine obtained from patients with renal disorders. *Clin Nephrol, 49*, 214-220.

Koibuchi, N., Konno, R., Matsuzaki, S., Ohtake, H., Niwa, A. & Yamaoka, S. (1995). Localization of D-amino acid oxidase mRNA in the mouse kidney and the effect of testosterone treatment. *Histochem Cell Biol, 104*, 349-355.

Konno, R. (1998a). Methods for the detection of D-amino-acid oxidase. *Biological Procedures Online, 1*, 27-31.

Konno, R. (1998b). Rat D-amino-acid oxidase cDNA: rat D-amino-acid oxidase as an intermediate form between mouse and other mammalian D-amino-acid oxidases. *Biochim Biophys Acta, 1395*, 165-170.

Konno, R. (2001). Assignment of D-amino-acid oxidase gene to a human and a mouse chromosome. *Amino Acids, 20*, 401-408.

Konno, R. (2003). Rat cerebral serine racemase: amino acid deletion and truncation at carboxy terminus. *Neurosci Lett, 349*, 111-114.

Konno, R. & Yasumura, Y. (1983). Mouse mutant deficient in D-amino acid oxidase activity. *Genetics, 103*, 277-285.

Konno, R. & Yasumura, Y. (1984). Involvement of D-amino-acid oxidase in D-amino acid utilization in the mouse. *J Nutr, 114*, 1617-1621.

Konno, R. & Yasumura, Y. (1988). A simple and rapid method to screen for mutant mice lacking D-amino-acid oxidase activity. *Lab Anim Sci, 38*, 292-295.

Konno, R., Isobe, K., Niwa, A. & Yasumura, Y. (1988a). Lack of D-amino-acid oxidase activity causes a specific renal aminoaciduria in the mouse. *Biochim Biophys Acta, 967*, 382-390.

Konno, R., Isobe, K., Niwa, A. & Yasumura, Y. (1988b). Excessive urinary excretion of methionine in mutant mice lacking D-amino-acid oxidase activity. *Metabolism, 37*, 1139-1142.

Konno, R., Nagata, Y., Niwa, A. & Yasumura, Y. (1989a). Spontaneous excretion of D-alanine in urine in mutant mice lacking D-amino-acid oxidase. *Biochem J, 261*, 285-287.

Konno, R., Niwa, A. & Yasumura, Y. (1989b). Linkage of the *Dao-1* gene for D-amino-acid oxidase to the *Pgm-1* gene for phosphoglucomutase-1 on the mouse chromosome 5. *Jpn J Genet, 64*, 341-345.

Konno, R., Niwa, A. & Yasumura, Y. (1990). Intestinal bacterial origin of D-alanine in urine of mutant mice lacking D-amino-acid oxidase. *Biochem J, 268*, 263-265.

Konno, R., Ikeda, M., Niwa, A. & Yasumura, Y. (1991a). D-Aminoaciduria in mutant mice lacking D-amino-acid oxidase activity. *Amino Acids, 1*, 239-246.

Konno, R., Yamamoto, K., Niwa, A. & Yasumura, Y. (1991b). Presence of D-amino-acid oxidase protein in mutant mice lacking D-amino-acid oxidase activity. *Int J Biochem, 23*, 1301-1305.

Konno, R., Oowada, T., Ozaki, A., Iida, T., Niwa, A., Yasumura, Y. & Mizutani, T. (1993). Origin of D-alanine present in urine of mutant mice lacking D-amino-acid oxidase activity. *Am J Physiol, 265*, G699-G703.

Konno, R., Sasaki, M., Shinohara, M., Enami, J. & Niwa, A. (1996). Mouse D-amino-acid oxidase gene: restriction fragment length polymorphism among mouse strains. *Amino Acids, 10*, 133-143.

Konno, R., Sasaki, M., Asakura, S., Fukui, K., Enami, J. & Niwa, A. (1997). D-Amino-acid oxidase is not present in the mouse liver. *Biochim Biophys Acta, 1335*, 173-181.

Konno, R., Ikeda, M., Yamaguchi, K., Ueda, Y. & Niwa, A. (2000). Nephrotoxicity of D-propargylglycine in mice. *Arch Toxicol, 74*, 473-479.

Krebs, H. A. (1935). CXCVII. Metabolism of amino-acids. III. Deamination of amino-acids. *Biochem J, 29*, 1620-1644.

Kun, E., Dechary, J. M. & Pitot, H. C. (1954). The oxidation of glycolic acid by a liver enzyme. *J Biol Chem, 210*, 269-280.

Lehmann, W. D., Theobald, N., Fischer, R. & Heinrich, H. C. (1983). Stereospecificity of phenylalanine plasma kinetics and hydroxylation in man following oral application of a stable isotope-labelled pseudo-racemic mixture of L- and D-phenylalanine. *Clin Chim Acta, 128*, 181-198.

Man, E. H. & Bada, J. L. (1987). Dietary D-amino acids. *Ann Rev Nutr, 7*, 209-225.

Matsui, T., Sekiguchi, M., Hashimoto, A., Tomita, U., Nishikawa, T. & Wada, K. (1995). Functional comparison of D-serine and glycine in rodents: the effect on cloned NMDA receptors and the extracellular concentration. *J Neurochem, 65*, 454-458.

Mattevi, A., Vanoni, M. A., Todone, F., Rizzi, M., Teplyakov, A., Coda, A., Bolongnesi, M. & Curti, B. (1996). Crystal structure of D-amino acid oxidase: a case of active site mirror-image convergent evolution with flavocytochrome $b_2$. *Proc Natl Acad Sci USA, 93*, 7496-7501.

Meister, A. (1965) *Biochemistry of the Amino Acids* (2nd edition) New York: Academic Press

Mizutani, H., Miyahara, I., Hirotsu, K., Nishina, Y., Shiga, K., Setoyama, C. & Miura, R. (1996). Three-dimensional structure of porcine kidney D-amino acid oxidase at 3.0 Å resolution. *J Biochem, 120*, 14-17.

Molla, G., Motteran, L., Piubelli, L., Pilone, M. S. & Pollegioni, L. (2003). Regulation of D-amino acid oxidase expression in the yeast *Rhodotorula gracilis*. *Yeast, 20*, 1061-1069.

Momoi, K., Fukui, K., Watanabe, F. & Miyake, Y. (1988). Molecular cloning and sequence analysis of cDNA encoding human kidney D-amino acid oxidase. *FEBS Lett, 238*, 180-184.

Momoi, K., Fukui, K., Tada, M. & Miyake, Y. (1990). Gene expression of D-amino acid oxidase in rabbit kidney. *J Biochem, 108*, 406-413.

Morehead, R. P., Poe, W. D., Williams, J. O. & Lazenby, M. E. (1946). The influence of age and species on the nephrotoxic action of DL-serine. *Am J Pathol, 22*, 658.

Moreno, S., Nardacci, R., Cimini, A. & Cerù, M. P. (1999). Immunocytochemical localization of D-amino acid oxidase in rat brain. *J Neurocytol, 28*, 169-185.

Morikawa, A., Hamase, K., Inoue, T., Konno, R., Niwa, A. & Zaitsu, K. (2001). Determination of free D-aspartic acid, D-serine and D-alanine in the brain of mutant mice lacking D-amino-acid oxidase activity. *J Chromatogr B, 757*, 119-125.

Mothet, J.-P., Parent, A. T., Wolosker, H., Brady, R. O., Jr., Linden, D. J., Ferris, C. D., Rogawski, M. A. & Snyder, S. H. (2000). D-Serine is an endogenous ligand for the glycine site of the *N*-methyl-D-aspartate receptor. *Proc Natl Acad Sci USA, 97*, 4926-4931.

Nagasaki, H., Yamada, R., Konno, R., Yasumura, Y. & Iwashima, A. (1990). D-Aspartate oxidase activity and D-aspartate content in a mutant mouse strain lacking D-amino acid oxidase. *Experientia, 46*, 468-470.

Nagata, Y. (1992). Involvement of D-amino acid oxidase in elimination of D-serine in mouse brain. *Experientia, 48*, 753-755.

Nagata, Y., Yamada, R., Nagasaki, H., Konno, R. & Yasumura, Y. (1991). Administration of D-alanine did not cause increase of D-amino acid oxidase activity in mice. *Experientia, 47*, 835-838.

Nagata, Y., Horiike, K. & Maeda, T. (1994). Distribution of free D-serine in vertebrate brains. *Brain Res, 634*, 291-295.

Nakajima, H., Ohta, M., Yamano, T. & Miyake, Y. (1981). The effect of D-C-propargylglycine and D-C-allylglycine on rats with special reference to D-amino acid oxidase. *Biomed Res, 2*, 154-165.

Negri, A., Ceciliani, F., Tedeschi, G., Simonic, T. & Ronchi, S. (1992). The primary structure of the flavoprotein D-aspartate oxidase from beef kidney. *J Biol Chem, 267*, 11865-11871.

Neidle, A. & Dunlop, D. S. (2002). Allosteric regulation of mouse brain serine racemase. *Neurochem Res, 27*, 1719-1724.

Perotti, M. E., Gavazzi, E., Trussardo, L., Malgaretti, N. & Curti, B. (1987). Immunoelectron microscopic localization of D-amino acid oxidase in rat kidney and liver. *Histochem J, 19*, 157-169.

Pilone, M. S. (2000). D-Amino acid oxidase: new findings. *Cell Mol Life Sci, 57*, 1732-1747.

Pollegioni, L., Diederichs, K., Molla, G., Umhau, S., Welte, W., Ghisla, S. & Pilone, M. S. (2002). Yeast D-amino acid oxidase: structural basis of its catalytic properties. *J Mol Biol, 324*, 535-546.

Ronchi, S., Minchiotti, L., Galliano, M., Curti, B., Swenson, R. P., Williams, C. H., Jr. & Massey, V. (1982). The primary structure of D-amino acid oxidase from pig kidney. II. Isolation and sequence of overlap peptides and the complete sequence. *J Biol Chem, 257*, 8824-8834.

Sarower, M. G., Okada, S. & Abe, H. (2003). Molecular characterization of D-amino acid oxidase from common carp *Cyprinus carpio* and its induction with exogenous free D-alanine. *Arch Biochem Biophys, 420*, 121-129.

Sasaki, M., Konno, R., Nishio, M., Niwa, A., Yasumura, Y. & Enami, J. (1992). A single-base-pair substitution abolishes D-amino-acid oxidase activity in the mouse. *Biochim Biophys Acta, 1139*, 315-318.

Sasamura, T., Matsuda, A. & Kokuba, Y. (2002). Determination of D-amino acid oxidase activity in tumour cells. *Ann Clin Biochem, 39*, 595-598.

Schell, M. J., Brady, R. O. J., Molliver, M. E. & Snyder, S. H. (1997). D-Serine as a neuromodulator: regional and developmental localizations in rat brain glia resemble NMDA receptors. *J Neurosci, 17*, 1604-1615.

Schell, M. J., Molliver, M. E. & Snyder, S. H. (1995). D-Serine, an endogenous synaptic modulator: localization to astrocytes and glutamate-stimulated release. *Proc Natl Acad Sci USA, 92*, 3948-3952.

Schieber, A., Brückner, H., Rupp-Classen, M., Specht, W., Nowitzki-Grimm, S. & Classen, H.-G. (1997). Evaluation of D-amino acid levels in rat by gas chromatography-selected ion monitoring mass spectrometry: no evidence for subacute toxicity of orally fed D-proline and D-aspartic acid. *J Chromatogr B, 691*, 1-12.

Setoyama, C. & Miura, R. (1997). Structural and functional characterization of the human brain D-aspartate oxidase. *J Biochem, 121*, 798-803.

Setoyama, C., Nishina, Y., Tamaoki, H., Mizutani, H., Miyahara, I., Hirotsu, K., Shiga, K. & Miura, R. (2002). Effects of hydrogen bonds in association with flavin and substrate in flavoenzyme D-amino acid oxidase. The catalytic and structural roles of Gly313 and Thr317. *J Biochem, 131*, 59-69.

Silbernagl, S., Völker, K. & Dantzler, W. H. (1999). D-Serine is reabsorbed in rat renal pars recta. *Am J Physiol, 276*, F857-F863.

Simonic, T., Duga, S., Negri, A., Tedeschi, G., Malcovati, M., Tenchini, M. L. & Ronchi, S. (1997). cDNA cloning and expression of the flavoprotein D-aspartate oxidase from bovine kidney cortex. *Biochem J, 322*, 729-735.

Snyder, S. H. & Kim, P. M. (2000). D-Amino acids as putative neurotransmitters: focus on D-serine. *Neurochem Res, 25*, 553-560.

Stevens, E. R., Esguerra, M., Kim, P. M., Newman, E. A., Snyder, S. H., Zahs, K. R. & Miller, R. F. (2003). D-Serine and serine racemase are present in the vertebrate retina and contribute to the physiological activation of NMDA receptors. *Proc Natl Acad Sci USA, 100*, 6789-6794.

Stříšovský, K., Jirásková, J., Bařinka, C., Majer, P., Rojas, C., Slusher, B. S. & Konvalinka, J. (2003). Mouse brain serine racemase catalyzes specific elimination of L-serine to pyruvate. *FEBS Lett, 535*, 44-48.

Tada, M., Fukui, K., Momoi, K. & Miyake, Y. (1990). Cloning and expression of a cDNA encoding mouse kidney D-amino acid oxidase. *Gene, 90*, 293-297.

Todone, F., Vanoni, M. A., Mozzarelli, A., Bolognesi, M., Coda, A., Curti, B. & Mattevi, A. (1997). Active site plasticity in D-amino acid oxidase: a crystallographic analysis. *Biochemistry, 36*, 5853-5860.

Tsai, G., Yang, P., Chung, L.-C., Lange, N. & Coyle, J. T. (1998). D-Serine added to antipsychotics for the treatment of schizophrenia. *Biol Psychiatry, 44*, 1081-1089.

Tsuchida, H., Yamamoto, N., Kajii, Y., Umino, A., Fukui, K. & Nishikawa, T. (2001). Cloning of a D-serine-regulated transcript *dsr-1* from the rat cerebral cortex. *Biochem Biophys Res Comm, 280*, 1189-1196.

Tsuchiya, M., Kurabayashi, A. & Konno, R. (2003). Hamster D-amino-acid oxidase cDNA: rodents lack the 27th amino acid residue in D-amino-acid oxidase. *Amino Acids, 24*, 223-226.

Uo, T., Yoshimura, T., Shimizu, S. & Esaki, N. (1998). Occurrence of pyridoxal 5'-phosphate-dependent serine racemase in silkworm, *Bombyx mori*. *Biochem Biophys Res Commun, 246*, 31-34.

Usuda, N., Yokota, S., Hashimoto, T. & Nagata, T. (1986). Immunocytochemical localization of D-amino acid oxidase in the central clear matrix of rat kidney peroxisomes. *J Histochem Cytochem, 34*, 1709-1718.

Usuda, N., Yokota, S., Ichikawa, R., Hashimoto, T. & Nagata, T. (1991). Immunoelectron microscopic study of a new D-amino acid oxidase-immunoreactive subcompartment in rat liver peroxisomes. *J Histochem Cytochem, 39*, 95-102.

Wake, K., Yamazaki, H., Hanzawa, S., Konno, R., Sakio, H., Niwa, A. & Hori, Y. (2001). Exaggerated responses to chronic nociceptive stimuli and enhancement of *N*-methyl-D-aspartate receptor-mediated synaptic transmission in mutant mice lacking D-amino-acid oxidase. *Neurosci Lett, 297*, 25-28.

Wako, K., Ma, N., Shiroyama, T. & Semba, R. (1995). Glial uptake of intracerebroventricularly injected D-serine in the rat brain: an immunocytochemical study. *Neurosci Lett, 185*, 171-174.

Watanabe, T., Motomura, Y. & Suga, T. (1978). A new colorimetric determination of D-amino acid oxidase and urate oxidase activity. *Anal Biochem, 86*, 310-315.

Wolosker, H., Sheth, K. N., Takahashi, M., Mothet, J.-P., Brady, R. O, Jr., Ferris, C. D. & Snyder, S. H. (1999a). Purification of serine racemase: biosynthesis of the neuromodulator D-serine. *Proc Natl Acad Sci USA, 96*, 721-725.

Wolosker, H., Blackshaw, S. & Snyder, S. H. (1999b). Serine racemase: a glial enzyme synthesizing D-serine to regulate glutamate-*N*-methyl-D-aspartate neurotransmission. *Proc Natl Acad Sci USA, 96*, 13409-13414.

Wood, P. L., Hawkinson, J. E. & Goodnough, D. B. (1996). Formation of D-serine from L-phosphoserine in brain synaptosomes. *J Neurochem, 67*, 1485-1490.

Yamamoto, N., Tomita, U., Umino, A. & Nishikawa, T. (2001). Uptake of D-serine by synaptosomal P2 fraction isolated from rat brain. *Synapse, 42*, 84-86.

Yasuda, E., Ma, N. & Semba, R. (2001). Immunohistochemical evidences for localization and production of D-serine in some neurons in the rat brain. *Neurosci Lett, 299*, 162-164.

Yoda, N., Konno, R. & Nagashima, S. (2001). Establishment of a rat hepatoma-derived cell line proliferating in D-phenylalanine medium and expressing D-amino-acid oxidase. *Amino Acids, 20*, 175-185.

In: Trends in DNA Research
Editor: Corey R. Woods, pp. 123-144
ISBN 1-59454-635-5


*Chapter VI*

# Characterizations of DNA Primary Sequences

***Chun Li*** [1,2,*] ***and Zhi-fu Wang*** [1]
[1] Department of Mathematics
[2] Institute of Mathematics, Bohai University, Jinzhou, P. R. China

## Abstract

The rapid growth of DNA sequences data in various DNA databanks leads to that the sequence analysis becomes more important but still difficult task. To extract the information stored in the sequence consisting of letters on the alphabet $\Sigma=\{A, C, G, T\}$, many mathematical schemes have been proposed. They are mainly the graphical representations, the numerical characterizations, and the method of characteristic sequences. Among them, the graphical representation gives a visual and hence qualitative characterization of DNA sequences, while the numerical characterization provides a quantitative analysis of data on DNA sequences. The method of characteristic sequences is a new approach based on the idea of the coarse-grained description of the DNA primary sequence.

**Key words:** DNA, Sequence analysis, Graphical representations, Numerical characterizations, Characteristic sequences.

## Introduction

The 21st century is a century of life sciences. Its milestone is the Human Genome Project (*HGP*) which lasts 13 years. With the completion of the genome projects of human, mouse

* Corresponding author e-mail: lchlmb@yahoo.com.cn. Corresponding author address: Department of Mathematics, Bohai University, Jinzhou 121000, P. R. China.

and rat one after another, and the development of some similar projects for other species, the number of known DNA sequences is accumulated at an exponential rate with respect to time. A DNA sequence is usually expressed in terms of a series of four letters A, G, C, and T, which denote the four nucleic acid bases: adenine, guanine, cytosine, and thymine, respectively. This expression is called a Letter Sequence Representation (LSR) or a DNA primary sequence. The LSR of a DNA sequence is necessary for the information storage and some statistical analysis. However, it is extremely difficult to recognize, to remember and to compare the sequences intuitively, especially for the long sequences, based merely on the LSR of the sequences. To overcome the drawback of the LSR, many kinds of mathematical approaches, such as graphical representations, numerical characterizations, and the method of characteristic sequences, have been proposed. In the next sections, we detailedly introduce these schemes and selectively show their applications to the sequence analysis.

# The Graphical Representations of DNA Primary Sequences

The graphical representations of DNA primary sequences provide a simple way of viewing, sorting and comparing various gene structures. Existing examples mainly include 2- and 3-dimensional graphical representations.

## The 2-D Graphical Representations

The 2-D graphical representation was first proposed by Gates [1] in 1986, and rediscovered independently by Nandy [2, 3] and Leong and Mogenthaler [4]. Their method is based on choosing the four directions, ($-x$), ($+x$), ($-y$), and ($+y$), in ($x$, $y$) coordinate system to represent the content of the four bases in DNA sequences. The method essentially consists of plotting a point corresponding to a base by moving one unit in the positive or negative $x$- or $y$-axis depending on the defined association of a base with a cardinal direction; the cumulative plot of such points produces a graph that corresponds to the sequence of bases in the gene fragment under consideration. For example, in Nandy's approach, they construct a symmetric purine-pyrimidine graph on the Cartesian coordinate system, with purines on the $x$-axis (A along the negative $x$-axis, G along the positive $x$-axis) and pyrimidines on the $y$-axis (T in the negative $y$-direction, C in the positive) and plot the sequence structure using the following rule: Starting from the origin, plot a point for each succeeding nucleotide with a displacement depending on the type of nucleotide: for adenine the displacement is one unit in the negative $x$-direction; for cytosine, it is one unit up from the current position in the positive $y$-direction; for guanine the displacement is one unit to the right, and for thymine it is one unit down. Then the total displacement up to say the $n$-th nucleotide on the sequence will be represented by a point on the purine-pyrimidine plot where the displacement along the $x$-axis represents the difference of total number of G's and the total number of A's up to this $n$-th nucleotide, and the displacement along the $y$-axis represents the cumulative difference between the C's and T's up to that nucleotide. As $n$ goes from 1 to $N$, the total sequence length, there will be a progression of points on the map generating a graphical representation of the pattern of composition of a DNA sequence which can be expected to differ for different sequences.[1] Take the segment ATGGTGCACC, the first 10 bases of the exon-1 of human *beta*-globin gene for an example; we show its 2-D Nandy's graph in Figure 1.

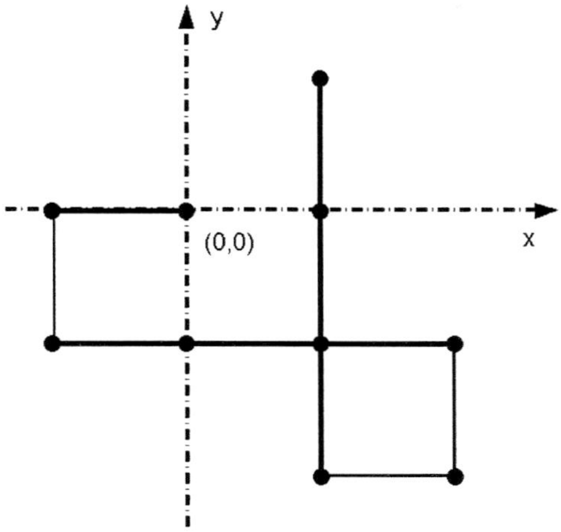


Figure 1. The 2-D Nandy's graphical representation of the sequence ATGGTGCACC

The technique for graphical representation of gene sequences on a two-dimensional Cartesian coordinate system has been shown to be able to provide considerable information on new global sequence patterns and homologies, repeated structures, relative base abundances, probable evolutionary paths and evolutionary divergences, and has been found to be a useful tool for revealing differences in the features of intron and exon segments of eukaryotic sequences (details, see [2, 5, 6]).

However, based merely on the 2-D graph in Figure 1, we are, contrarily, unable to judge which sequence it corresponds to, the sequences segment ATGGTGCACC, or ATGGGTACCC? The reason for it is simple: there is an intersection in the 2-D curve. Meanwhile, it is not difficult to find that the sequences AG, AGA, AGAG, AGAGA, AGAGAG... have the same graphical representation in the view of the geometry. Usually, we say a curve/graph has degeneracy if it corresponds to $m$ ($m$>1) DNA primary sequences. Clearly, the degeneracies in the 2-D Nandy's curve are consistent with that there are multiple edges or circuits from the point of view of the graph theory.[2]

To reduce the degeneracy, Guo, Randic and Basak [8] improved the representation by a modification of the directions of vectors assigned to the four bases. In detail, they design four

---

[1] In the Gates axes system, one would move one unit in the positive $x$-direction for a C, along the positive $y$-direction for a T, the negative $x$-direction for a G and the negative $y$-direction for an A. While the Leong and Morgenthaler axes system associates A with positive $x$-direction, T with positive $y$-direction, C with negative $x$-direction, and G with negative $y$-direction.

[2] Nandy and Nandy [7] demonstrated that such degeneracies exist only in certain restrictive cases, and in particular, for the SNP genes and several other classes of DNA sequences, the DNA descriptors arising from those 2-D graphical representations are unique and thus can be used to systematically index and tabulate such sequences.

special vectors in Cartesian ($x$, $y$) coordinate system to represent the four nucleic acid bases as follows:

$$(-1, 1/d) \rightarrow \text{A},$$
$$(1/d, -1) \rightarrow \text{T},$$
$$(+1, 1/d) \rightarrow \text{G},$$
$$(1/d, +1) \rightarrow \text{C},$$

where $d$ is a positive integer (see Figure 2)

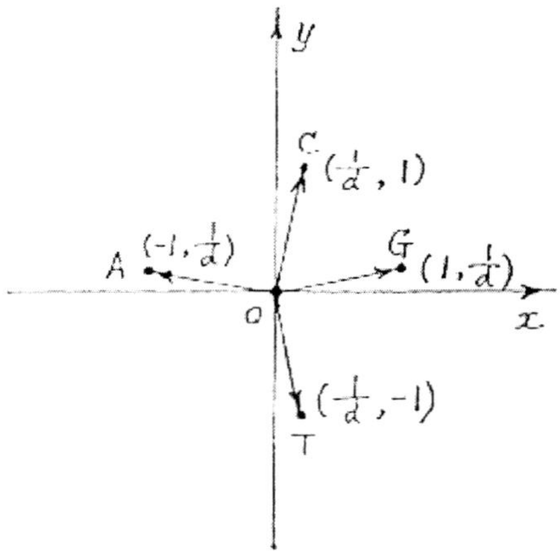


Figure. 2. The directions designed to represent four nucleotides A, G, C and T, respectively.

It is showed that the minimum length of a circuit in this graphical representation of a DNA sequence is equal to either $4d$ for an even integer $d$ or $2d$ for an odd integer $d$. This implies that, if $d$ is a greater even number, the 2-D Guo's graphical representation of a DNA primary sequence has lower degeneracy.

For the same purpose, Wu, Liew, Yan and Yang [9] introduced the DB-Curve (Dual-Base Curve) by displaying two of the four bases at a time on a plan. The idea is that if a sequence exhibits interesting visual features, this should also be visible in the sub-sequence consisting of two of the bases. Two sequences that are similar should also have their similarity reflected in their sub-sequences consisting of two of the bases. Take the AC DB-Curve as an example. A vector with start point (0, 0) and end point (+1, +1) is defined to correspond to base A, a vector with start point (0, 0) and end point (-1, +1) to base C, a vector with start point (0, 0) and end point (0, +1) to bases T and G (see Figure 3). If define the starting point as (0, 0), a DNA sequence can be mapped to a 2D-coordinate system by a cumulative plot of the bases in the sequence using the above notation. The AC DB-Curve emphasizes the relationship between A and C bases and makes their visualization simple and clear. Other five DB-Curves (AG, AT, TC, CG and TG DB-curves) can be obtained in a similar way.

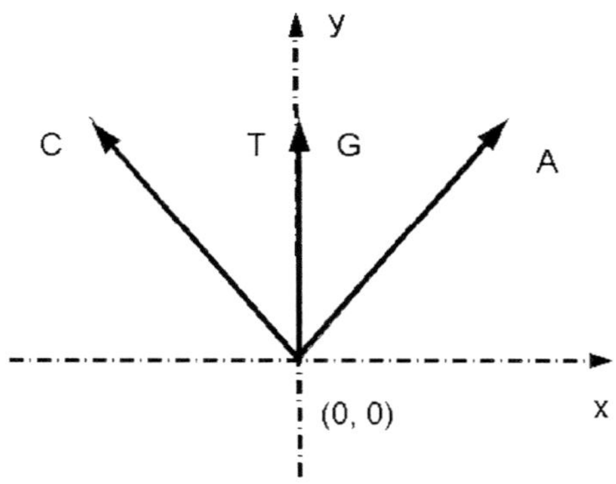


Figure. 3. The directions designed to represent four nucleotides A, G, C and T, respectively.

Similarly, Yau et al. [10], as shown in Figure 4, constructed a pyrimidine-purine graph on two quadrants of the Cartesian coordinate system, with pyrimidines (T and C) in the first quadrant and purines (A and G) in the fourth quadrant. The unit vectors representing four nucleotides A, G, C and T are as follows:

$$(\frac{1}{2},-\frac{\sqrt{3}}{2}) \to A,\ (\frac{\sqrt{3}}{2},-\frac{1}{2}) \to G,\ (\frac{\sqrt{3}}{2},\frac{1}{2}) \to C,\ (\frac{1}{2},\frac{\sqrt{3}}{2}) \to T$$

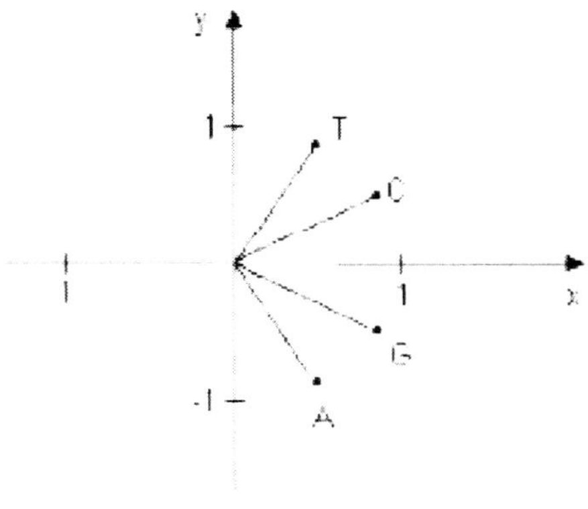


Figure 4. The unit vectors designed to represent four nucleotides A, G, C and T, respectively.

Instead of considering the four directions in the cartesian $(x, y)$ coordinate system, as the above authors did, Randic, Vracko, Lers, and Plavsic [11] proposed a 'four horizontal lines' scheme. They draw four horizontal lines separated by unit distances, on which dots (rectangles) representing the bases constituting the considered sequence are placed. The representation requires first to associate the four types of bases with the four horizontal lines. They label the lines in the order as the types of bases appear for the first time in the sequence, which is A, T, G, and C. The consecutive bases along the horizontal axes are placed at unit displacement. Connecting adjacent dots, one can obtain a zigzag like curve (see Figure 5).

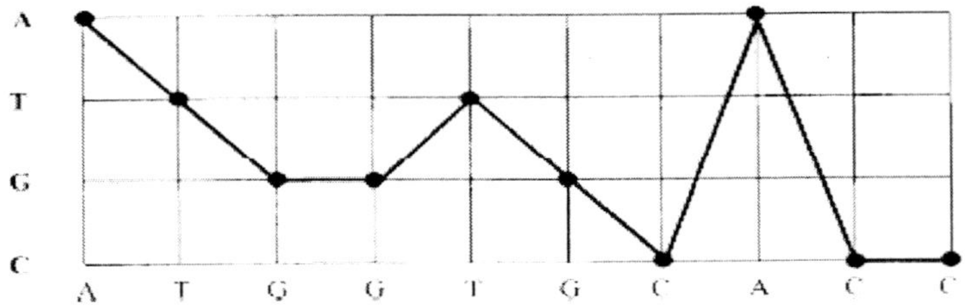


Figure 5. The graphical representation of the sequence ATGGTGCACC. The rectangles (dots) denote the bases making up the sequence.

Clearly, the 'four horizontal lines' graphical representation as well as the 2-D Yau's graph has avoided the problem of degeneracy because the two curves have a monotonic increasing characteristic. However, by each of the two methods, one might need 12 different curves to represent the same DNA sequence.

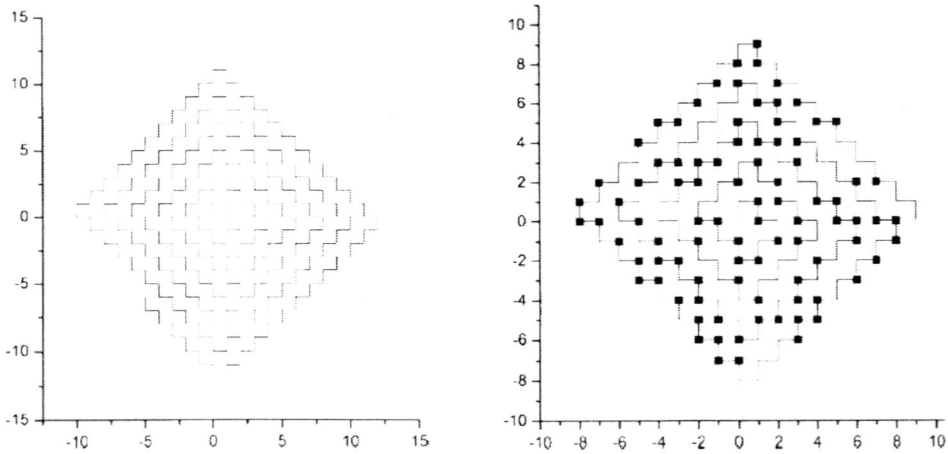


(a) (b)

Figure 6. (a). The zigzag spiral representing the 'worm' curve used as template for construction of compact graphical representation of DNA. (b). The compact graphical representation of the first exon of human *beta*-globin gene.

In addition, a 'compact' 2-D graphical representation [12, 13] was proposed which does not require considerable space yet is capable to fully preserve local characteristics of the DNA sequence. The compact representation is based on using a compact mathematical curve as a template on which nucleic bases are displayed. Once a template curve (which has vertices at equidistant positions) has been selected, one can assign to its vertices nucleotide bases according to a selected protocol. Randic et al. first represent the four nucleic acid bases adenine (A), guanine (G), cytosine (C), and tyrosine (T) by binary labels 00, 01, 10 and 11, respectively. Thus, to be recorded, each base requires a pair of adjacent vertices on the template curve. In the next step they replace the binary labels 00, 01, 10 and 11 by white and black circles as their graphical equivalents. Retaining only the black circles corresponding to digit '1' of the binary codes, a compact graphical representation of DNA is then obtained.

Figure 6(a) is a compact curve which appears as a zigzag spiral, known in the artificial intelligence literature as the 'worm' curve. The curve is defined by a moving point over the Cartesian coordinate grid following the rule: At every site go to the right (*i.e.*, make a $90^o$ turn), if the move does not brings you to a place already visited, otherwise go to the left (*i.e.*, make a $270^o$ turn). The worm curve represents a path of a simple robot but is of interest to us for two reasons: (1) it does not intersect itself, and (2) it uses as little space as possible. Fig 6.(b) shows graphically the DNA sequence for the first exon of human *beta*-globin gene by suppressing 'zeros' and depicting only 'ones' as dark spots.

## The 3-D Graphical Representations

Twenty years ago, Hamori and Ruskin [14] first used a 3-D H curve to represent a DNA sequence. The basic rule for constructing H-curve is to move one unit along one of four directions (NW, NE, SE, and SW) representing four bases in *xy*-plane, and one for each unit in the *z*-direction. We can see that the H-curve needs more space since its height always equals the length of the DNA sequence considered.

The Z curve, proposed by Zhang and Zhang [15] in 1994, is another 3-D graphical representation of DNA sequence. It is composed of a series of nodes $P_0$, $P_1$, $P_2$, . . . , $P_N$, whose coordinates $x_n$, $y_n$ and $z_n$ ($n$ = 0, 1, 2, . . . , $N$, where $N$ is the length of the DNA sequence being studied) are *uniquely* determined by the Z-transform of DNA sequence

$$\begin{cases} x_n = (A_n + G_n) - (C_n + T_n), \\ y_n = (A_n + C_n) - (G_n + T_n), \\ z_n = (A_n + T_n) - (C_n + G_n), \end{cases}$$

$$x_n, y_n, z_n \in [-N, N], n = 0,1,2,\cdots,N,$$

where $A_n$, $C_n$, $G_n$ and $T_n$ ($n$ = 1, 2, . . . , $N$) are the *cumulative* occurrence numbers of A, C, G and T, respectively, in the subsequence from the 1st base to the $n$-th base in the sequence, and $A_0 = C_0 = G_0 = T_0 = 0$. The Z curve is defined as the connection of the nodes $P_0$, $P_1$, $P_2$, . . . , $P_N$ one by one sequentially with straight lines. The three components of the Z curve, $x_n$, $y_n$ and $z_n$, display the distributions of purine/pyrimidine (R/Y), amino/keto (M/K) and weak-H bond/strong-H bond (W/S) bases along the sequence, respectively. In the subsequence constituted from the 1st base to the $n$-th bases of the sequence, when purine bases (A/G) are in excess of pyrimidine bases (C/T), $x_n > 0$, otherwise, $x_n < 0$, and when the numbers of purine (A/G) and pyrimidine bases (C/T) are identical, $x_n = 0$. Similarly, when amino bases (A/C) are in excess of keto bases (G/T), $y_n > 0$, otherwise, $y_n < 0$, when the numbers of amino (A/C) and keto bases (C/T) are identical, $y_n = 0$. Finally, when weak H-bond bases (A/T) are in excess of strong H-bond bases (G/C), $z_n > 0$, otherwise, $z_n < 0$, and when the numbers of (A/T) and (G/C) bases are identical, $z_n = 0$. It is not difficult to find that the H curve, the 2-D

Gates' curve, and then the 2-D curves of Nandy, Leong and Mogenthaler are all the special forms of the Z curves [15].

The major advantages of the Z curve methodology are as follows (see http://tubic.tju.edu.cn/zcurve/) :

- The Z curve is a basic visualization tool of DNA sequences. Global and local features of DNA sequence can be grasped quickly in a perceivable form.
- Comparison two or more genome sequences can be performed by viewing their Z curves in a computer screen or a piece of paper only, regardless of how long these sequences are.
- The Z curve method is extremely simple, and very easy to use.
- The introduction of the Z curve initiates a new way to analyze DNA sequences using a geometric approach.

Nevertheless, the Z curve also has degeneracy due to overlapping and crossing of the curve representing DNA with itself. Take the sequence AGCTAGCTAGCT for an example, its corresponding nodes $P_0, P_1, P_2, \ldots, P_{12}$ are as follows:

(0,0,0), (1,1,1), (2,0,0), (1,1,-1), (0,0,0) , (1,1,1), (2,0,0), (1,1,-1), (0,0,0), ...

which leads to that there is an overlapped circuit in the Z curve of the sequence segment. In spite of this, the Z curve has proved most useful (see [15-25]).

In 2000, Randic, Vracko, Nandy and Basak [26] generalized directly the 2-D graphical representations of Gates, Nandy, Leong and Mogenthaler to a 3-D graphical representation. They place the origin of the Cartesian $(x, y, z)$ coordinate system in the center of a cube so that the four corners of the cube have the coordinates (+1, -1, -1), (-1, +1, -1), (-1, -1, +1), (+1, +1, +1), and assign four nucleic bases as follows:

(+1, -1, -1) → A
(-1, +1, -1) → G
(-1, -1, +1) → C
(+1, +1, +1) → T

To obtain the graphical representation of a DNA sequence, they start from the origin, and move in $(x, y, z)$ space in the direction that the above assignment dictates. Hence a DNA sequence corresponds to a curve in the 3-D space. However, as pointed out by Guo, Randic and Basak [8], the degeneracy of the 3-D graphical representation of DNA sequences is still higher. For example, the sequences AGTCA, AGTCAG, AGTCAGT, AGTCAGTC, ..., AGTCAGTCAGTCAG, ... have the same graphical representation (see Figure 7 ).

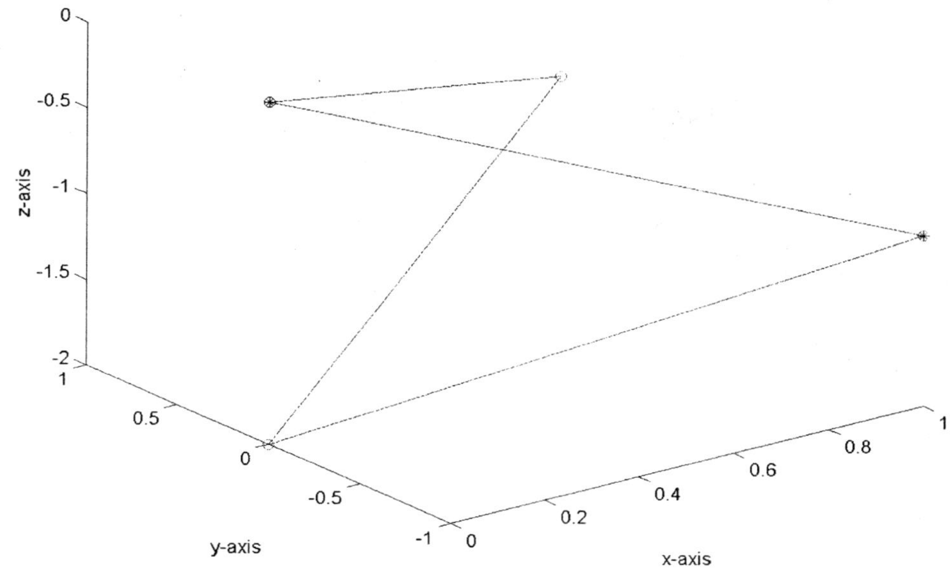


Figure 7. The 3-D graphical representations of sequences AGTCA, AGTCAG, AGTCAGT, AGTCAGTC, ..., AGTCAGTCAGTCAG.

By assigning four special directions to the four nucleic bases:

(1,0,0) →A,
(0,1,0) →C,
(0,0,1) →G,
(1,1,1) →T,

Li and Wang [27] introduced another 3-D graphical representation of DNA sequences which no longer intersects or overlaps itself and therefore the degeneracy is avoided completely. Especially, a repetitive fragment of a DNA sequence will have corresponding repetitive sections (see Figure 8). In addition, If it is projected to planes, three different 2-D graphical representations are obtained, each of which can be regarded as an analog of a DB-Curve in [9] with a bit 'degeneracy'. For example, its 2-D projection on the *xy*-plane is similar to the AC DB-Curve. In Figure 9, we show the projection on *xy*-plane and the AC DB-curve of the sequence segment GAGTTTTATCGCTTCCATGA, the first 20 bases of the gene called coliphage phiX174 (GI: 9626372). It is easy to see that the two graphs almost have the same 'trace'.

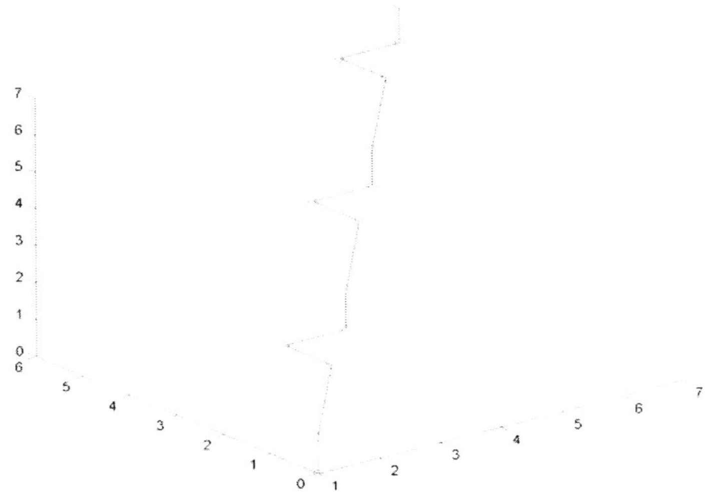


Figure 8. The 3-D Li's curve of sequence AGTCAGTCAGTCAG.

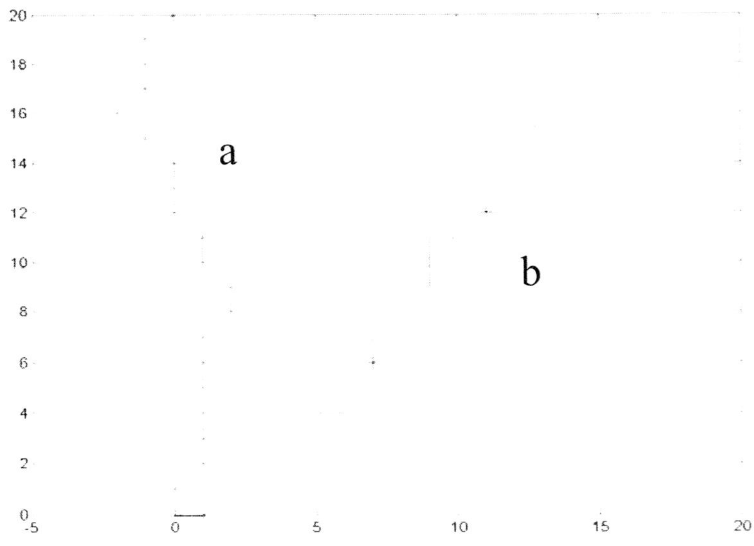


Figure 9. AC DB-curve and the projection on *xy*-plane of the 3-D curve of the sequence segment GAGTTTTATCGCTTCCATGA, the first 20 bases of the gene coliphage phiX174. (a) AC DB-curve, (b) the projection on xy-plane of the Li's 3-D curve.

Similar to Randic's 'four horizontal lines' graph, Yuan et al. [28] proposed a 'four pillar' graphical representation for DNA sequences. In detail, let $X = X_1 X_2 \cdots$ be an arbitrary DNA primary sequence. Then they define a map $\phi$, which maps $X$ into a plot set. Explicitly, $\phi(X) = \phi(X_1)\phi(X_2)\ldots$, where

$$\phi(X_i) = \begin{cases} (-1,0,i) & if \ X_i = A \\ (+1,0,i) & if \ X_i = G \\ (0,-1,i) & if \ X_i = T \\ (0,+1,i) & if \ X_i = C \end{cases}$$

Thus a DNA primary sequence is transformed into a series of points. By connecting these points, a 3-D curve is obtained. Figure 10 shows the 'four pillar' graphical representation of the sequence ATGGTGCACC. As we see the 'four pillar' graphical representation also has no degeneracy, and it requires no display of 2-D projections for visualization and analysis.

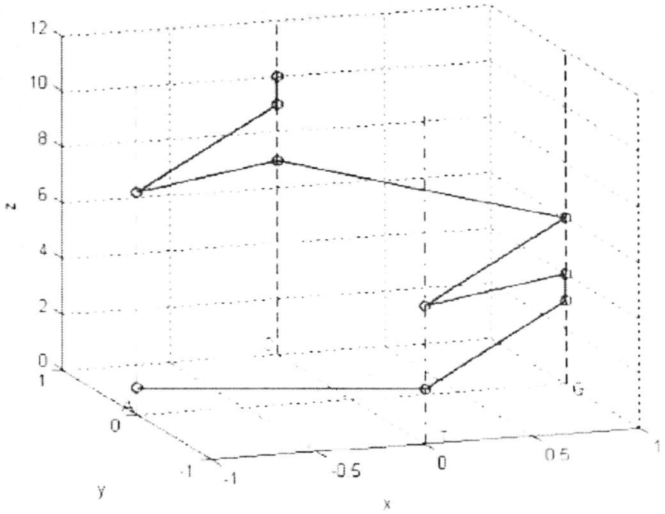


Figure 10. The 'four pillar' graphical representation of the sequence ATGGTGCACC

## The Higher Dimensional Representations

In addition to the 2-D and 3-D curves of DNA sequences, we can see some higher dimensional representations in the literature.

For example, The G curve reported by Hamori and Ruskin [14, 29] was generated in a virtual 5-D space whose coordinates are each assigned to the four DNA nucleotides and to an integer characterizing the position of nucleotide on a DNA sequence. Also, in [30], a 4-D representation of DNA primary sequences was considered by assigning to each of the four bases A, T, G, C directions along the four orthogonal coordinate axes:

A (1, 0, 0, 0)
T (0, 1, 0, 0)
G (0, 0, 1, 0)
C (0, 0, 0, 1)

Because the four directions of 4-D space are fully equivalent, the above selection is equivalent to any other permutation of labels and directions, hence this selection should not be viewed as introducing any arbitrary decision to influence numerical analysis.

It is easy to see both of the two curves no longer intersect or overlap themselves. However, the important advantage of graphical visualization of DNA sequences has been lost.

# Numerical Characterization

Quantitative methods have provided us with an important technique on the aspect of mathematical analysis of large volume genomic DNA sequence data, and several authors have proposed different schemes to numerically characterize DNA sequences.

## The *ED, GD, PD, M/M, L/L* (also *Q*) Matrices [11, 31]

The matrices mentioned in this subsection are all constructed from graphical representations of DNA sequences. For our purpose, suppose that we have transformed a given DNA sequence into a curve embedded in a $k$-dimensional space.

### *The Euclidean-distance matrix ED*

The $(i, j)$-matrix element is defined to be the Euclidean-distance between vertices $i$ and $j$ of the curve in the $k$-dimensional space:

$$[ED]_{ij} = \sqrt{(x_{i1} - x_{j1})^2 + (x_{i2} - x_{j2})^2 + \cdots + (x_{ik} - x_{jk})^2}$$

### *The Graph Theoretical Distance matrix GD*

The $(i, j)$-matrix element is defined to be the graph theoretical distance between vertices $i$ and $j$ of the curve:

$$[GD]_{ij} = |j-i|$$

### *The Path-distance Matrix PD*

The $(i, j)$-matrix element is defined to be the sum of the Euclidean-distances between the adjacent vertices along the path from vertex $i$ to vertex $j$:

$$[PD]_{ji} = [PD]_{ij} = [ED]_{i,i+1} + [ED]_{i+1,i+2} + \ldots + [ED]_{j-1,j},\ i < j;$$

$$[PD]_{ii} = 0$$

### *The M/M matrix*

The $(i, j)$-matrix element is defined to be the quotient of the corresponding elements of the *ED* matrix and the *GD* matrix:

$[M/M]_{ij} = [ED]_{ij}/[GD]_{ij}$ , $i \neq j$;
$[M/M]_{ii} = 0$.

### *The L/L matrix*

The (*i, j*)-matrix element of *L/L* is defined to be the quotient of the corresponding elements of the *ED* matrix and the *PD* matrix:

$[L/L]_{ij} = [ED]_{ij}/[PD]_{ij}$ , $i \neq j$;
$[L/L]_{ii} = 0$.

### *The 'Higher Order' matrix* $^{n}M$

From a matrix *M*, one can construct $^{n}M$ matrices, namely the product of Hadammard multiplication of the *M* matrix by itself *n*-times, from which additional structurally related descriptors can be derived. Take the $^{n}L/^{n}L$ matrix for an example, its definition is

$$\left[{}^{n}L/{}^{n}L\right]_{ij} = \left([L/L]_{ij}\right)^{n}.$$

It is easy to see that for any two vertices *i* and *j* ($i \neq j$), the inequality $0 < [ED]_{ij} \leq [PD]_{ij}$ always holds, and hence $0 < [{}^{n}L/{}^{n}L]_{ij} \leq 1$. Particularly, as *n* approaches infinity, the matrix sequences $\{{}^{n}L/{}^{n}L\}$ converge to the binary matrix ${}^{b}L/{}^{b}L$, which can be obtained from the *L/L* matrix by substitution of all elements <1 with zero.

## The *S/S* Matrix [32]

Instead of depending on a geometrical representation of the primary DNA sequences, which involves, to a greater or lesser degree, an arbitrary convention when the assignment of the direction in *xy*-plane, or *xyz*-space are selected for the four nucleic bases A, G, C and T, the *S/S* matrix is obtained by regarding the primary sequence of DNA itself as an input.

Let $X = X_1X_2X_3 \cdots X_N$ be an arbitrary DNA primary sequence. Then the (*i, j*)-matrix element of *S/S* is defined as follows:

$$s_{ji} = s_{ij} = \frac{n_{ij}}{j-i}, \ (j>i);$$

$$s_{ii} = 0.$$

Here $n_{ij}$ is the number of the nucleic bases same as the one corresponding to $X_j$ in the subsequence $X_{i+1}X_{i+2} \cdots X_j$.

As we have seen, the orders of the *S/S* matrices are always equal to the lengths of the DNA sequences considered. Therefore, lengthy sequences will produce very 'large' matrices. In the same paper, Randic et al. proposed a $4 \times 4$ condensed matrix the elements of which are all extracted from the larger matrix (details, see [32]).

## Frequency Matrix

The matrices in this subsection are mainly formed based on the statistics, and usually obtained from a DNA primary sequence directly. For our purpose, we suppose $\sum=\{A,G,C,T\}$.

- Denoting by $f_L$ ($L \in \Sigma$) the frequency of the solo nucleotide $L$ in the sequence under study, *i.e.*, $f_L = \frac{L_N}{N}$, where $L_N$ is the total number of $L$'s, and $N$ the length of the sequence, we obtain a matrix (vector) with 4 entries:
  - $(f_A, f_G, f_C, f_T)_{1\times 4}$
- If we denote the frequency of the dinucleotide $XY$ in the sequence by $f_{XY}$ $(XY \in \Sigma \times \Sigma)$, then we obtain a matrix (vector) with $4^2$ entries:
  - $(f_{AA}, f_{AG}, f_{AC}, f_{AT}, f_{GA}, \ldots, f_{TT})_{1\times 16}$

Generally speaking, if we consider the frequencies of the sub-strings with $k$ bases (the $k$-words) in the DNA sequence under study, then we will obtain a matrix/vector with $4^k$ entries. For convenience, we denote such a matrix/vector by $\mathbf{F}^k$. The applications of $\mathbf{F}^3$ can be found in Randic et al. [33].

Moreover, based on the first two kinds of frequencies, Karlin and Burge [34] defined the dinucleotide odds ratios as follows:

$$\rho_{XY} = {f_{XY}}\Big/{f_X f_Y},$$

which is a useful parameter for DNA sequence analysis.

Similarly, He [35] introduced the sieve ratio of trinucleotide and relative abundance profiles of $k$-words to characterize and compare coding sequences. The relative abundance profile of a $k$-word $a_1 a_2 \cdots a_k$ is defined as:

$$\rho_{a_1 a_2 \cdots a_k} = \frac{f_{a_1 a_2 \cdots a_k} \times f_{a_2 \cdots a_{k-1}}}{f_{a_1 a_2 \cdots a_{k-1}} \times f_{a_2 \cdots a_k}}$$

Clearly, both the dinucleotide odds ratio and the sieve ratio of trinucleotide are the special cases of the relative abundance profile.

## Nandy's 2-D Graph Descriptors [2, 5, 6, 7]

Once the Nandy's 2-D graphical representation of a DNA sequence is given, one can define a weighted mean of the $x$-coordinate values of the representative points as $u_x = \sum x_i / N$ and mean $y$-coordinate value as $u_y = \sum y_i / N$, where $N$ represents the total length of the DNA sequence. This gives rise to the first of the two sets of DNA descriptors - the vector representing the weighted mean point with magnitude

$$g_R = [u_x^2 + u_y^2]^{\frac{1}{2}}$$

which is defined as the graph radius, subtending with the $x$-axis an angle $\theta$ given by

$$\theta = u_y / u_x .$$

Since the $g_R$ is very sensitive to small changes in the sequence, when comparing two DNA sequences one can compare the distance between their two graph radii and obtain the second set of DNA descriptors - the difference in $g_R$, $\Delta_R$, between two graphs whose magnitude is defined as

$$\Delta_R = [(u_x - u_x') + (u_y - u_y')]^{1/2},$$

with a corresponding defining angle

$$\delta\theta_R = \theta - \theta'$$

The two sets of DNA descriptors have been proved to be suitable for using it as a parameter to index SNP related genes and toxicity levels of various chemicals that induce changes in DNAs [7].

## DNA Sequence Invariant

The approach characterizing DNA sequences by ordered sets of invariants was proposed by Randic et al. (see [26, 33, 36]). A sequence invariant is a real number that is independent of the labels (bases) A, G, C, and T. For example, the length of a sequence is an invariant.

But the length cannot capture the main information of the sequence considered, so it is regarded as a trivial invariant. Other invariants include the *average matrix element, average row sum, the leading eigenvalue,* and *Wiener number* of a matrix associated with the DNA sequence, etc.. Among them, the Wiener number *W*(*G*), in fact, is a widely used descriptor of molecular topology, which is introduced by Wiener as the path number for saturated acyclic hydrocarbons, and the path number is defined as the number of the bonds between all pairs of non-hydrogen atoms in the molecule. In other words, the Wiener number *W*(*G*) of a structure *G* is equal to the half-sum of entries of the graph theoretical distance matrix $D(G) = (d_{ij})$ of *G* [32, 37]:

$$W(G) = \frac{1}{2}\sum_{ij} d_{ij}$$

We can see the *average matrix element*, *average row sum* and *Wiener number* are related to each other, and differ from one another only by normalization factor. The leading eigenvalue is an important matrix invariant and is proved to be highly effective for characterization of DNA sequences. However, a trouble with which we must faced is the calculation of the eigenvalue will become more and more difficult with the order of the matrix large. According to the Frobenius-Perron theorem of linear algebra the largest and the smallest row sums represent the upper and the lower bounds, respectively, for the leading eigenvalue of a symmetric matrix. Therefore, the *average row sum* has an advantage, particularly when the individual row sums do not differ widely, because it may suggest an approximate value for the leading eigenvalue of the matrix.

Furthermore, if $M = (a_{ij})$ is an $n \times n$ nonnegative real symmetric matrix with diagonal entries zero, *i.e.*, $a_{ij} = a_{ji} \geq 0$ and $a_{ii} = 0$ for $i,j = 1,2,\ldots,n$ (in fact, the *ED*, *GD*, *PD*, *M/M*, *L/L*, and *S/S* matrices associated with DNA sequences are all such matrices), then the following inequality always holds [38-41]:

$$\frac{1}{n}\|M\|_{m1} \leq \lambda_1 \leq \sqrt{\frac{n-1}{n}}\|M\|_F \tag{a}$$

where $\lambda_1$ is the leading eigenvalue of *M*, $\|M\|_F \equiv \sqrt{\sum_{i,j}|a_{ij}|^2} = \sqrt{tr(M^T M)}$, and $\|M\|_{m1} \equiv \sum_{i,j}|a_{ij}|$. It is interesting that $\frac{1}{n}\|M\|_{m1}$ is just the *average row sum* of matrix *M*, therefore (a) implies that the *average row sum* of *M* is the lower bound of the leading eigenvalue of *M*. Moreover, both the bounds in (a) are attained [38], so their arithmetic average, $\chi = \chi(M) = \frac{1}{2}\left(\frac{1}{n}\|M\|_{m1} + \sqrt{\frac{n-1}{n}}\|M\|_F\right)$, can be regarded as an approximate value of the leading eigenvalue of $M$. It is just in this sense that $\chi(M)$ is called as 'ALE-index' [38], and in the same paper, the authors point out that the ALE-index $\chi$ and the leading eigenvalue $\lambda_1$ play a similar role in sequence analysis, while $\chi$ is simpler for calculation. So one can use the ALE-index as an alternative invariant to characterize DNA sequences.

## Entropy-like Measures of DNA Sequences

### *The Shannon Entropy of a DNA Sequence*

Based on the Shannon's information theory, several authors [17, 42, 43] proposed the Shannon entropy $H$ of a DNA sequence, which is defined by

$$H = -\sum_{i=1}^{4} p_i \log_2 p_i ,$$

where $p_i$ ($i = 1,2,3,4$) are the occurrence frequencies of bases A, C, G, and T in the sequence, respectively.

It is not difficult to see the entropy of a DNA sequence defined as above only reflects the statistical structure of the sequence considered. In fact, for a sequence, the order relation among its elements as well as the elements themselves includes important information. To reflect the information on elements of the sequence and, especially, the order relation among them, Li and Wang [44] introduced a so-called normalized relative-entropy.

### *The Normalized Relative-entropy of a Sequence of Numbers*

Suppose $S = x_1, x_2, \cdots, x_k$ is a finite sequence of non-negative real numbers. Then we can regard it as an indexed multiset $S = \{x_1, x_2, \cdots, x_k\}$. To reflect the order of $x_i$'s one can construct a *chain* $D(S) = \{x_1, x_1 + x_2, \cdots, x_1 + x_2 + \cdots + x_k\} \hat{=} \{X_1, X_2, \cdots, X_k\}$. It is easily seen that the indexed multiset $S$ and the *chain* $D(S)$ determine each other uniquely.

As a discrete set, $D(S)$ can be thought of as a point partition of the global value $W$ of some properties into $k$ terms $W = X_1 + X_2 + \cdots + X_k$. For this partition, one can construct a discrete probability distribution $P = (p_1, p_2, \cdots, p_k)$, where $\sum_i p_i = 1$, and $p_i = X_i / W$.

The Shannon entropy of the discrete probability distribution is calculated by the formula [45, 46]:

$$H(S) = H(D(S), P) = -\sum_{i=1}^{k} p_i \log_2 p_i . \qquad \text{(b)}$$

The entropy defined by Eq.(b) has the following properties:

(1) For any discrete probability distribution with $k$ components, the entropy is in the closed interval $[0, \log_2 k]$.

(2) Suppose $S=\{x_1, x_2, \cdots, x_k\}$. Then $H(mS)=H(S)$ for all $m>0$. Where $mS=\{mx_1, mx_2, \cdots, mx_k\}$.

(3) Let $\varepsilon_1 = \{1,0,\cdots,0,0\}, \varepsilon_k = \{0,0,\cdots 0,1\}$, and $S=\{x_1, x_2, \cdots, x_k\}$. Then $H(S)=0$ if and only if there exists an $m>0$ such that $S=m\varepsilon_k$; while $H(S)=\log_2 k$ if and only if there exists an $m>0$ such that $S=m\varepsilon_1$.

(4) Let $\varepsilon_1 = \{1,0,\cdots,0,0\}$, $\varepsilon_2 = \{0,1,\cdots,0,0\}$, $\cdots$, $\varepsilon_k = \{0,0,\cdots 0,1\}$. Then $H(\varepsilon_1) > H(\varepsilon_2) > \cdots > H(\varepsilon_k)$. Notice that $\varepsilon_i$ corresponds to the position $i$ in a sequence of number, therefore the property (4) implies that the entropy can reflect the degree of the importance of the position in a sequence.

Furthermore, the relative-entropy, namely the relative-ratio of entropy is calculated by

$$\eta(S) = H(S) \big/ H_{\max} \tag{c}$$

where $H_{\max} = \log_2 k$.

If we treat an indexed $k$-multiset $S$ as a vector in a $k$-dimensional space (in fact, one often does so), then by property (2) and Eq.(c), we have: two $k$-multisets (as vectors) in the same direction have the same relative-entropy. Meanwhile, in the view of the geometry, in order to precisely describe a vector $v=(x_1, x_2, \cdots, x_k)$ in a $k$-dimensional space, one needs two parameters: one is the direction and the other is its length, $\|v\| = \sqrt{\sum_i x_i^2}$. Taking into account of this point and the influence caused by different dimensions/sizes of vectors/sets, Li and Wang [44] describe a finite sequence of non-negative real numbers $S=\{x_1, x_2, \cdots, x_k\}$ by

$$h_L = h_L(S) = \frac{H(S) \big/ H_{\max}}{\|S\| \big/ \sqrt{k}} \tag{d}$$

where $\|S\| = \sqrt{\sum_i x_i^2}$. For convenience, they call $h_L(S)$ (or $h_L$) normalized relative-entropy of the sequence $S$.

As well known, DNA sequences consisting of letters on an alphabet $\Sigma=\{A, C, G, T\}$ can be transformed into some sequences of numbers in a certain way. Once the sequences of non-negative real numbers are given, their normalized relative-entropies (see Eq.(d)) can be calculated to characterize the corresponding DNA sequence. It is showed that this is a useful method for sequence analysis.

# Characteristic Sequences

In 2002, He and Wang [47] introduced another representation for DNA sequences, which is based on the idea of the coarse-grained description of the DNA primary sequence. Usually, the four nucleic acid bases A, G, C and T can be divided into two classes according to their chemical structures, *i.e.* purine R = {A, G} and pyrimidine Y = {C, T}. The bases can be also divided into another two classes, amino group M = {A, C} and keto group K = {G, T}. In addition, the division can be made according to the strength of the hydrogen bond, *i.e.* weak H-bonds W ={A, T} and strong H-bonds S ={G, C}. For a given DNA sequence $X = \mathrm{x}_1\mathrm{x}_2\cdots$, He and Wang define three homomorphic maps $\phi_i\,(i = 1, 2, 3)$ by $\phi_i(\mathrm{X}) = \phi_i(\mathrm{x}_1)\phi_i(\mathrm{x}_2)\cdots$, where

$$\phi_1(\mathrm{x}_j) = \begin{cases} 1 & if\ \mathrm{x}_j \in R \\ 0 & if\ \mathrm{x}_j \in Y \end{cases} \quad (j = 1, 2, \cdots)$$

$$\phi_2(\mathrm{x}_j) = \begin{cases} 1 & if\ \mathrm{x}_j \in M \\ 0 & if\ \mathrm{x}_j \in K \end{cases} \quad (j = 1, 2, \cdots)$$

$$\phi_3(\mathrm{x}_j) = \begin{cases} 1 & if\ \mathrm{x}_j \in W \\ 0 & if\ \mathrm{x}_j \in S \end{cases} \quad (j = 1, 2, \cdots)$$

Thus, three (0,1)-sequences corresponding to the same DNA sequence can be obtained, which are named the (R, Y)-, (M, K)-, and (W, S)-characteristic sequences of the DNA sequence, respectively [47]. (They are called as logic sequences of the DNA sequence in [48].) The following mathematical theorem says that the three characteristic sequences give all information of the primary sequence.

*Theorem 1.* A DNA primary sequence is uniquely determined by any pair of its three characteristic sequences (details, see [47]).

As pointed out in [47, 49], comparing characteristic sequences, one can get some information that cannot be obtained from the direct comparison of DNA primary sequences and observe some special nature in species from different aspect. Although some information may be lost in characteristic sequences, one can focus his attention on the information of his interest. This is the advantage of the approach.

For each characteristic sequence, Li and Wang [50] associated a ${}^bL/{}^bL$ matrix by giving a 2-D graphical representation based on 'two horizontal lines' and calculated the 'pseudo-trace' of the ${}^bL/{}^bL$ matrix. They thus produced a 3-component vector whose components are the 'pseudo-traces' of three ${}^bL/{}^bL$ matrices corresponding to the three characteristic sequences. It is showed that the 3-component vector results in more simple characterizations and comparisons among the coding sequences of exon 1 of *beta*-globin gene of eleven different species.

# Conclusion

As we know, the sequence analysis is an important task in bioinformatics. Of all need to do, the characterization of a DNA sequence is the first step. In this chapter, we introduce three kinds of mathematical schemes for characterizing DNA sequences: the graphical representations, the numerical characterizations, and the method of characteristic sequences. Each of which has advantages and disadvantages. How to develop these approaches and how to combine them efficiently will be interesting problems.

# Acknowledgements

This work was supported in part by the Educational Department of Liaoning Province of China (20040013).

# References

[1] Gates, M.A., A simple way to look at DNA, *J. Theor. Biol.*, 119 (1986), 319.

[2] Nandy, A., A new graphical representation and analysis of DNA sequence structure: I. Methodology and application to globin genes. *Curr. Sci.* 66 (1994), 309.

[3] Nandy, A., Graphical representation of long DNA sequences. *Curr. Sci.* 66 (1994), 821.

[4] Leong, P.M.; Mogenthaler, S., Random walk and gap plots of DNA sequences, *Comput. Appl. Biosci.* 12 (1995), 503.

[5] Nandy, A.; Nandy, P., Graphical analysis of DNA sequences structure: II. Relative abundance of nucleotides in DNAs, gene evolution and duplication, *Curr, Sci.*, 68 (1995), 75.

[6] Nandy, A., Graphical analysis of DNA sequence structure: III. Indication of evolutionary distinctions and characteristics of introns and exons, *Curr. Sci.*, 70 (1996), 661

[7] Nandy, A.; Nandy, P., On the uniqueness of quantitative DNA difference descriptors in 2D graphical representation models, *Chem. Phys. Lett.* 368 (2003), 102.

[8] Guo, X.F.; Randic, M.; Basak, S.C., A novel 2-D graphical representation of DNA sequences of low degeneracy, *Chem. Phys. Lett.* 350 (2001), 106.

[9] Wu, Y.H.; Liew, A.W.; Yan, H.; Yang, M., DB-Curve: a novel 2D method of DNA sequence visualization and representation, *Chem. Phys. Lett.* 367 (2003), 170.

[10] Yau, S.S.T.; Wang, J.; Niknejad, A.; Lu, C.; Jin, N.; Ho, Y.K., DNA sequence representation without degeneracy. *Nucleic Acids Res.* 31 (2003), 12.

[11] Randic, M.; Vracko, M.; Lers, N.; Plavsic, D., Novel 2-D graphical representation of DNA sequences and their numerical characterization, *Chem. Phys. Lett.* 368 (2003), 1.

[12] Randic, M.; Vracko, M.; Zupan, J.; Novic M., Compact 2-D graphical representation of DNA. *Chem. Phys. Lett*, 373 (2003), 558.

[13] Randic, M., Graphical representations of DNA as 2-D map. *Chem. Phys. Lett*, 386 (2003), 468.

[14] Hamori, E.; Ruskin, J., H curves, a novel method of representation of nucleotide series especially suited for long DNA sequences. *J. Biol. Chem.*, 258 (1983), 1318.

[15] Zhang, R.; Zhang, C.T., Z curves, an intuitive tool for visualizing and analyzing DNA sequences. *J. Biomol. Struc. Dyn.*, 11 (1994), 767

[16] Zhang, C.T.; Chou, K.C., A Graphic Approach to Analyzing Codon Usage in 1562 E.Coli Protein Coding Sequences. *J. Mol. Biol.* 238 (1994), 1-8.

[17] Zhang, C.T., A Symmetrical Theory of DNA Sequences. *J. Theor. Biol.* 187 (1997), 297-306.

[18] Zhang, C.T.; Lin, Z.S.; Yan, M.; Zhang, R., A Novel Approach To Distinguish Between Intron-Containing and Intronless Genes Based on the Format of Z Curves. *J. Theor. Biol.* 192 (1998), 467-473.

[19] Yan, M.; Lin, Z.S.; Zhang, C.T., A New Fourier Transform Approach for Protein Coding Measure Based on the Format of the Z Curve. *Bioinformatics* 14 (1998), 685-690.

[20] Zhang, C.T.; Wang, J., Recognition of Protein Coding Genes in the Yeast Genome at Better Than 95% Accuracy Based on the Z curve. *Nucleic Acids Res.* 28 (2000), 2804-2814.

[21] Wang, J.; Zhang, C.T., Identification of protein-coding genes in the genome of Vibrio cholerae with more than 98% accuracy using occurrence frequencies of single nucleotides. *Eur. J. Biochem.* 268 (2001), 4261-4268.

[22] Zhang, C.T.; Wang, J.; Zhang, R., Using a Euclid distance discriminant method to find protein coding genes in the yeast genome. *Compt. Chem.* 26 (2002), 195-206.

[23] Guo, F.B.; Ou, H.Y.; Zhang, C.T., ZCURVE: a new system for recognizing protein-coding genes in bacterial and archaeal genomes. *Nucleic Acids Res.* 31 (2003), 1780-1789.

[24] Zhang, C.T.; Zhang, R.; Ou, H.Y., The Z curve database: a graphic representation of genome sequences. *Bioinformatics* 19 (2003), 593-599.

[25] Chen, L.L.; Zhang, C.T., Gene recognition from questionable ORFs in bacterial and archaeal genomes. *J. Biomol. Struct. Dyn.* 21 (2003), 99-110.

[26] Randic, M.; Vracko, M.; Nandy, A.; Basak, S.C., On 3-D graphical representation of DNA primary sequences and their numerical characterization. *J. Chem. Inf. Comput. Sci.* 40 (2000), 1235.

[27] Li, C.; Wang, J., On a 3-D representation of DNA primary sequences, *Comb. Chem. High T. Scr.*, 7 (2004), 23.

[28] Yuan, C.X.; Liao, B.; Wang, T.M., New 3D graphical representation of DNA sequences and their numerical characterization. *Chem. Phys. Lett.* 379 (2003), 412.

[29] Hamori, E. BioTechniques, 7 (1989), 710.
[30] Randic, M.; Balaban, A.T., On A Four-Dimensional Representation of DNA Primary Sequences. *J. Chem. Inf. Comput. Sci.* 43 (2003), 532.
[31] Randic, M.; Vracko, M.; Lers, N.; Plavsic, D., Analysis of similarity/dissimilarity of DNA sequences based on novel 2-D graphical representation, *Chem. Phys. Lett.* 371 (2003), 202.
[32] Randic, M., On characterization of DNA primary sequences by a condensed matrix. *Chem. Phys. Lett.* 317 (2000), 29.
[33] Randic, M.; Guo, X.F.; Basak, S.C., On the characterization of DNA primary sequences by triplet of nucleic acid bases. *J. Chem. Inf. Comput. Sci.* 41 (2001), 619.
[34] Karlin, S.; Burge, C., Dinucleotide relative abundance extremes: a genomic signature. *Trends in Genetics,* 11 (1995), 283.
[35] He, P-an, The Sieve Ratio for Characterization and Similarity Analysis of DNA Sequences. *Comb. Chem. & High T. Scr.* 8 (2005), 449-453.
[36] Randic, M.; Vracko, M., On the similarity of DNA primary sequences. *J. Chem. Inf. Comput. Sci.* 40 (2000), 599-606.
[37] Li X.; Li Z.; Hu, M., A novel set of Wiener indices. *J. Mol. Graph. Model.* 22 (2003) 161.
[38] Li, C.; Wang, J., New Invariant of DNA Sequences. *J. Chem. Inf. Model.* 45 (2005), 115-120.
[39] London, D. Inequalities in quadratic forms. *Duke Math. J.* 33 (1966), 511-522.
[40] Shrock, R.; Tsai, S.-H., Upper and lower bounds for ground state entropy of antiferromagnetic Potts models. *Phys. Rev. E.* 55 (1997), 6791-6794.
[41] Biggs, N. *Algebraic Graph Theory,* 1st ed. Cambridge, England: Cambridge University Press; 1974.
[42] Li, W.; Galvan-Bernaola, P.; Haghighi, F.; Grosse, I., Applications of recursive segmentation to the analysis of DNA sequences. *Computers and chemistry.* 26 (2002), 491.
[43] He, P-an; Wang, J., Some properties for DNA curve. *J. Phys. A: Math. Gen.* 37 (2004) 7135–7142.
[44] Li, C.; Wang, J., Relative Entropy of DNA and its Application, *Physica A: Statistical Mechanics and its Applications*, 347 (2005), 465-471.
[45] Guiasu, S., *Information Theory with Application,* McGraw–Hill; 1997.
[46] Martin, N.F.G.; England, J.W., *Mathematical theory of entropy*, Addison-Wesley Pub. Co.; 1981.
[47] He, P-an; Wang, J., Characteristic sequences for DNA primary sequence. *J. Chem. Inf. Comput. Sci.,* 42, (2002) 1080.
[48] Liu, Y., The numerical characterization and similarity analysis of DNA primary sequences, *Internet Electron. J. Mol. Des.* 1 (2002), 675.
[49] He, P-an; Wang, J., Numerical Characterization of DNA Primary Sequence, *Internet Electron. J. Mol. Des.*, 1, (2002) 668.
[50] Li, C.; Wang, J., Numerical characterization and similarity analysis of DNA sequences based on 2-D graphical representation of the characteristic sequences, *Comb. Chem. & High T. Scr.,* 6 (2003) 795.

In: Trends in DNA Research
Editor: Corey R. Woods, pp. 145-163
ISBN 1-59454-635-5


*Chapter VII*

# Impact of Polymeric Surfaces on Oligonucleotides Immobilization for Lab-on-a-Chip Applications

***Duy K. Pham*, Chee Kai Chan** and Elena P. Ivanova****
*Swinburne University of Technology, Hawthorn, Victoria, Australia
** Department of Genetics, La Trobe University, Bundoora, Victoria, Australia

## Abstract

The impact of three polymeric surfaces, i.e. carboxy-rich Poly(Styrene/Maleic Acid) (PSMA) and plasma treated cyclo-olefin-copolymer (COC) and polycarbonate (PC), on immobilization efficiency of oligonucleotides has been analyzed. The employment of different immobilization strategies, either chemically or via poly(L-lysine), have resulted in comparable high density (up to 16 x 1012 molecules/cm2) of amino-terminated oligonucleotides covalently immobilized on polymeric surfaces. The nature of DNA - poly(L-lysine) complexation is discussed. The relatively high immobilization efficiency coupled with the stable thermo-mechanical properties makes these polymers promising candidates for DNA-DNA hybridization assays integrated with PCR applications.

**Keywords***:* DNA immobilization, oligonucleotide, polysterine-derived polymeric surfaces, polycarbonate, cyclo-olefin copolymer, atomic force microscopy, oligonucleotide/poly(L-lysine) complexes.

## Introduction

The immobilization of DNA/oligonucleotides on solid supports is important in an array of bio- and nano-technological applications relying on DNA hybridization events (Fodor et al, 1991; Woolley et al, 1996; Cheng et al, 1996; Lin et al, 2000; Trau et al, 2002). Key requirements for the solid support of a fully integrated lab-on-a-chip device include high density of binding sites for immobilization of biomolecules, low intrinsic fluorescence, processability, and low surface roughness (Ivanova et al, 2002). While most of the initial demonstrations of these chips were achieved on glass, polymers are emerging as an attractive alternative due to their versatile chemical, physical and surface properties (Northrup et al, 1993, 1995). The cost of polymeric substrates may also be considerably less than that of glass or gold (Becker & Garner, 2000). In the category of cheap and easily processable materials, polymeric materials such as polystyrene, polycarbonate and cyclo-olefin copolymer are prime candidates.

DNA fragments, e.g. oligonucleotides, are attached on surfaces mainly through two very different approaches. First, the attachment is governed by purely physical process that is not specific and often reversible. Second, the DNA fragments are covalently bound to surfaces, a method that is entirely controlled by chemistry. There are various immobilization methods including (i) cross-linking to poly-L-lysine-coated microscope slide (Schema et al, 1995; Ivanova et al, 2004); (ii) photochemical immobilization, via photo reactive species to polymer surfaces (Koch et al, 2000); (iii) attachment of thiol-terminated oligonucleotides to thiol-functionalized surfaces (McGovern, Thompson, 1999; Boncheva et al, 1999; Rogers et al, 1999; Lenigk et al, 2001; Pham et al, 2002); (iv) attachment amino-functionalized surfaces (Zammatteo et al, 1996); (v) attachment of biotin-labeled oligonucleotides to streptavidin-functionalized surfaces (Belosludtsev et al, 2001).

The use of microarrays in large scale genomic analysis has led researchers to explore novel ways to immobilise short oligonucleotides onto microscopic arrays and to achieve efficient delivery of plasmid DNA. The electrostatic DNA condensation via various polyamino acids including poly(L-lysine) has only recently become widely in use as one on such methodologies due to its efficiency and low cost, even though the condensation of DNA has been of interest since the 1970s with the initial work focusing on how the DNA was packaged into heads of bacteriophages (Evdokimov et al, 1972; Laemmli 1975; Zama & Ichimura 1975; Bloomfield 1991). The notion that the condensation step is essential where the DNA is compacted by polycations like spermidine, protamine and poly-amino acids was crucial to evoke intensive research in this field (Bustamante & Rivetti 1996; Hansma et al, 1998; Chan et al, 2000). Initially poly(L-lysine) was used to conjugate a ligand via a heterobifunctional cross linkers to target the therapeutic plasmid DNA for cellular entry through the receptor-mediated- endocytosis in gene delivery experiments (Wu & Wu 1987; Wagner et al, 1991). It was later found that these DNA/poly(L-lysine) complexes also contribute in compaction of the plasmid DNA and the protection of the DNA from endonuclease digestion. A number of cell binding ligands including asialo-glycoproteins, transferrin, galactose, anti-CD3, anti-CD5 and ά-melanocyte stimulating hormone have been covalently linked to poly(L-lysine) for receptor mediated gene delivery. Membrane disrupting and nuclear targeting peptides can also been bioconjugated to poly(L-lysine) to aid endosomal escape and nuclear localization (see for review Wagner et al, 1998; Chan et al, 2002).

Polycations interact with DNA due to electrostatic attraction in which the positively charged groups of the polycation interacts with the negatively charged phosphate groups of the DNA. DNA condensation was investigated using a number of different techniques including agarose gel electrophoresis, circular dichroism, absorption spectroscopy, transmission electron microscopy, laser light scattering and more recently atomic force microscopy (AFM). The use of AFM has been particularly useful in obtaining three dimensional images at a sub-nanometre resolution at very near physiological conditions (Wolfert & Seymour 1996; Hansma et al. 1998; Dolan et al. 2001).

This study aims to evaluate: (i) the 'applicability' of three polymeric surfaces, i.e. carboxy-rich Poly(styrene/maleic acid) (PSMA) and plasma treated cyclo-olefin-copolymer (COC) and polycarbonate (PC), (ii) provide comparative analysis of immobilization efficiency of oligonucleotides after employment of poly(L-lysine) and/or N-hydroxy-succinimide (NHS) intermediate to covalently bind oligonucleotides, and (iii) describe the mechanism of condensation of poly(L-lysine) and DNA and properties of the complexes formed.

# Analysis of PSMA, Plasma Treated COC and PC Surfaces

## Analysis of PSMA Polymeric Surfaces

In our recent study of about 60 commercially available polymers, it appeared that only 22 polymers are in principle capable of DNA covalent binding (Ivanova et al, 2002). Of these polymers, two have amino groups or rather similar groups and high Tg's, hence a good prospect in their processability. However, it is questionable if the N substituted in an aromatic ring would allow an effective covalent binding of DNA. Another eight polymers have an anhydride group that in principle can be hydrolysed. Five polymers contain free carboxylic groups (possibly embedded in hydrogen bonding). These polymers were among the top candidates for direct covalent binding of DNA. Four polymers have either phenolic OH groups or alcoholic OH groups. Only the former polymers have high Tg's (for 2 methacrylates with alcoholic OH groups the Tg is unknown). These polymers were also good candidates for the covalent binding of DNA. In a class of its own, poly(tert-butyl methacrylate) (PtBuMA) does not have *per se* groups that can be used for the covalent binding of DNA. However, it is well known that PtBuMA can easily generate COOH groups upon irradiation with deep-UV light (or e-beam). Thus, polystyrene and/or its co-polymers, e.g. polystyrene-co-maleic acid (PSMA) has been selected for our studies. According to thermogravimetric analysis, PSMA had a valuable usage range of up to 130 °C, due to the change in transition state. Decarboxylation of the surface began to occur at around 160 °C. The transition from a solid to liquid state occurred at between 200-215 °C, thus positively defining the range of its thermal stability (Ivanova et al, 2002).

Surface topography of the PSMA surfaces was investigated using AFM. As can be seen from Figure 1, the overall roughness of the polymer surface is about few tens to 100 nm suggesting that it is rather flat with respect to biomolecular dimensions. PSMA is mildly hydrophilic with contact angle of 50°. One of the advantages of a hydrophilic surface is that it will favour an appropriate molecular approach path for covalent binding. As well, it has been argued (Chen et al., 2001) that hydrophobic surfaces can induce an unfavourable conformation of the adsorbed/covalently bound DNA strands with the bases facing down towards the surface. On the other hand, a hydrophilic surface may be unfavourable for microarray fabrication because "dots" of solution will spread out and limit the density of the microarray.

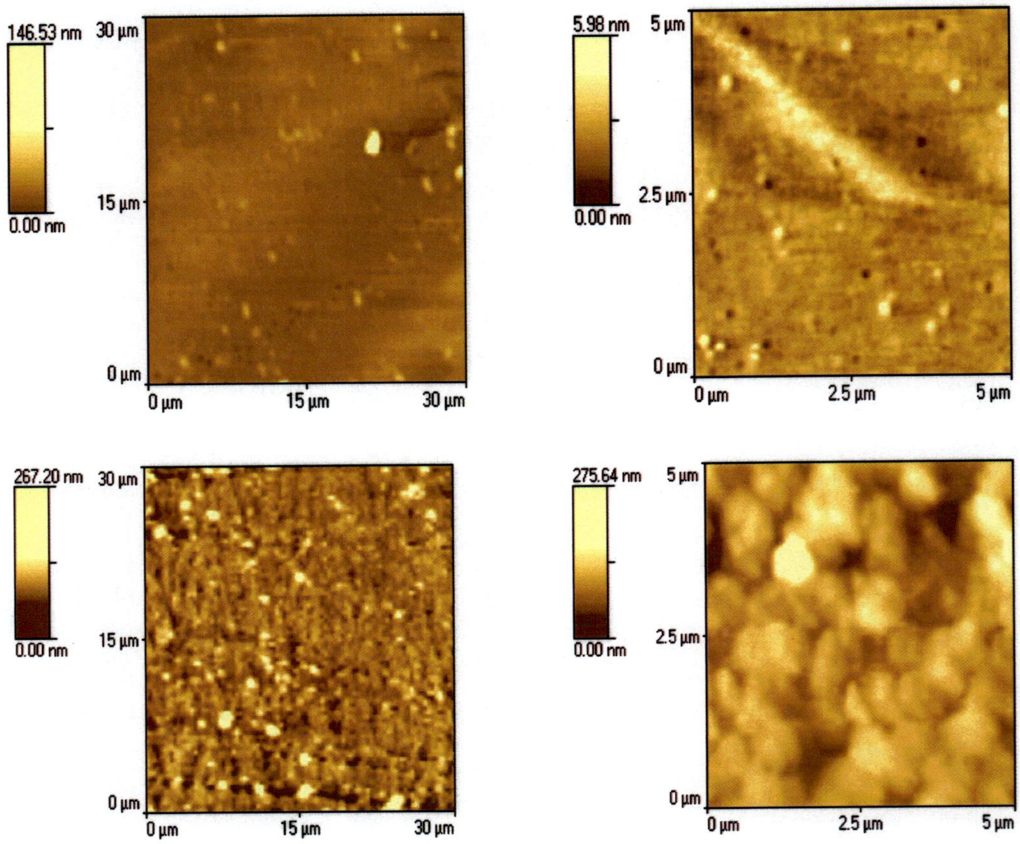


Figure 1. AFM characterization of the surface of PSMA (top) and oligonucleotides/PLy complexes bound onto PSMA via covalently attached PLy.

Although the carboxylic groups on the polymeric surface might have suffer reorientation towards the bulk of the polymer or suffer chemical alteration rendering them inoperable for DNA immobilisation, the XPS analysis confirmed the presence of carboxylic groups of the surface of PSMA as indicated by the presence of a distinct peak at 289.0 eV (Beamson and Briggs, 1992; Chastain, 1992, Ivanova et al, 2002).

Regarding the chemistry path, the COOH-based DNA immobilisation chemistry has the advantage of a high yield of reaction, but the disadvantage of a low surface concentration of DNA molecules (negative-negative electrostatic repulsion); whereas the $NH_2$-based chemistry provides a higher surface concentration (positive-negative electrostatic attraction). Therefore switching the immobilisation from a COOH-based to a $NH_2$-based chemistry may increase the efficiency of DNA immibilization (Ivanova et al, 2002, 2004).

## Analysis of Plasma-treated and NHS-functionalized Polymeric Surfaces

While PC is a common polymeric material, COC is a relatively new class of polymers, which is characterized by high moisture barrier, excellent thermal properties and bio-compatibility (Bergstrom et al, 2000; Chu et al, 2000; Delpech et al, 2001). Both of these polymers have glass transition temperatures (Tg) in excess of 150°C. In our study, COC and PC surfaces were first pre-treated with oxygen plasma, and then functionalized to create an N-hydroxy-succinimide (NHS) intermediate. The corresponding chemistry for the covalent binding of amino groups with carboxylic acid groups using NHS intermediate is described elsewhere (Lenigk et al, 2002).

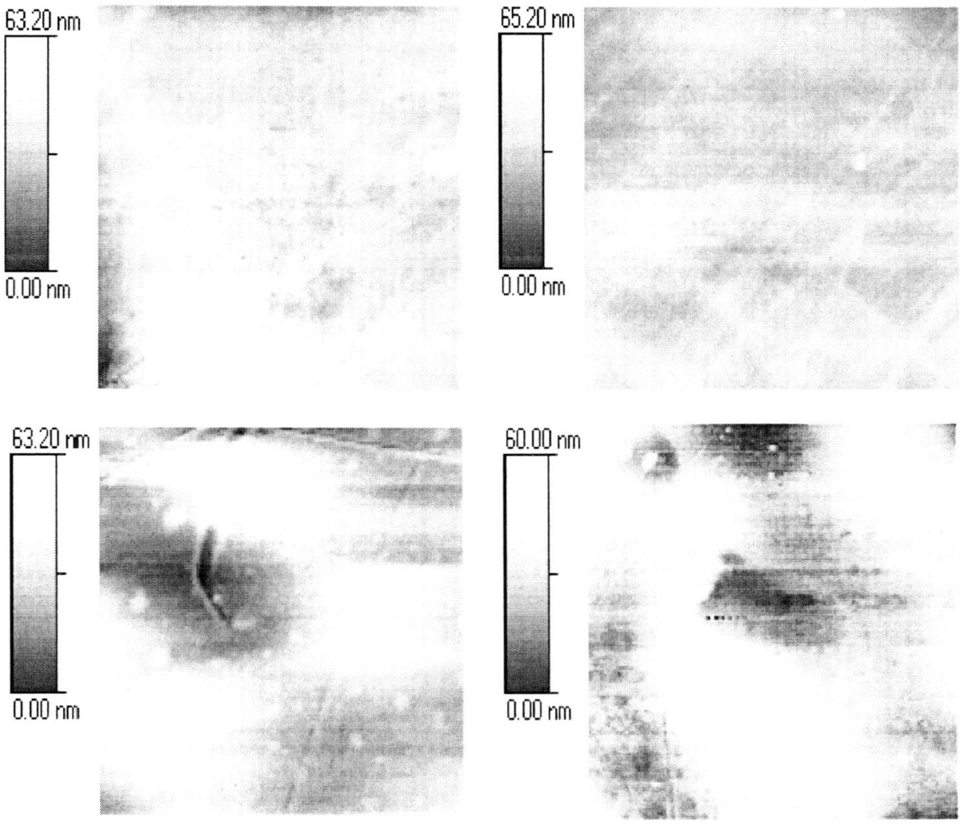


Figure 2. A. AFM topographical images of COC (left column) and PC (right column) samples: unmodified (top row); and 60 sec plasma treatment (bottom row). Surface area in view is 25 x 25 μm. B. AFM frictional force images of COC (left column) and PC (right column) samples: unmodified (top row); 60 sec plasma treatment (bottom row). Surface area in view is 25 x 25 μm. These lateral force images correspond to the respective topographical images in Figure 2a.

AFM imaging (Figure 2) shows that both COC and PC samples have relatively smooth textured surfaces. Plasma treatment, an essential step in the NHS-functionalization process, did not change the surface morphology significantly until after treatment for 60 sec or more. In particular, these surfaces exhibited either hill-like or crater-like features, on COC and PC, respectively, possibly due to arcing in the plasma reactor, a common problem for dielectric materials processed using plasma. The NHS-functionalization step appeared to smooth out the PC surface, whereas it did not change the surface roughness of the plasma-treated COC samples significantly. The RMS morphological roughness of the various COC and PC samples ranged from 4.57 nm to 8.62 nm (Pham et al, 2002).

When imaging under ambient conditions, the capillary condensation between the tip and sample surfaces reflects the relative degree of hydrophilicity and can be used as a basis for discriminating between hydrophobic and hydrophilic groups (Wilbur et al, 1995). It has been shown that when imaging in air, the frictional force is higher as the tip is scanned across a more hydrophilic region (Noy et al, 1995). The image contrast in a frictional force map is a measure of tip-to-surface friction, which reflects the surface chemical heterogeneity. Frictional force images of COC and PC samples before and after plasma treatment are shown in Figure 2. The corresponding RMS frictional roughness values are on a range of 0.85 - 4.04 nA. For both COC and PC surfaces, the plasma treatment increased the extent of surface chemical heterogeneity, with the COC surface becoming significantly more heterogeneous compared to the PC surface, as indicated by the respective increase in RMS frictional roughness (Pham et al, 2002). While the subsequent NHS-functionalization step appeared to smooth out the chemical contrast of the PC surface, it did not reduce the chemical heterogeneity of the plasma-treated COC surface significantly.

## Efficiency of Covalent Immobilization of Oligonucleoties

A 26 base-pair universal oligonucleotide primer 5' GTG GAT CAC CTG AGG TCA GGA GTT TC 3' corresponded to the '*alu* gene' (Qureshi et al, 1994; McCabe et al, 1999) was used in all experiments for attachment to selected polymeric surfaces via multifunctional cross linkers, 1-ethyl-3-(3-dimethylaminopropyl) carbodiimide hydrochloride (EDC) and *N*-hydroxysulfosuccinimide (NHSS) and/or poly(L-lysine) hydrobromide (PLy, MW 2000).

### Oligonucleotides Immobilization on PSMA Polymeric Surfaces

As PSMA presents surface chemical species fully oxidised, it exhibits a good chemical stability. However, linking DNA molecules, which are negatively charged due to phosphate groups, with the polymer carboxylic groups, which are also negatively charged, would ask for a charge/chemistry reversal from the negatively charged surface/DNA, either at the level of the polymer surface (as in the present study) or at the level of DNA molecules (via e.g. $NH_2$ groups at one end of the DNA/oligonucleotide molecule).

The results obtained indicated that the bound surface density (using EDC) was estimated as $0.3 \times 10^{10}$ molecules/mm$^2$ on PSMA, which is in fact lower the theoretical limit expected for a two-dimensional closest-packed array (Ivanova et al, 2002). Therefore employment of poly(L-lysine), a common surface modification poly-peptide, appeared to be a promising approach. In general, the polypeptides may adsorb onto a surface by an electrostatic attraction between the surface and charged amino acid residues. The mechanism and the expected strength of such adsorption is well documented (e.g. more recently by Yoon & Lenhoff, 1992; Hartley et al, 1993; Park & Healy, 2003). Method to create a charged surface was based on self-assembled monolayer of 11-mercaptoundecanoic acid on gold that can be negatively charged via the deprotonation of the carboxylic groups in the appropriate pH range (Jordan et al, 1994; Nuzzo et al, 1990; Jordan et al, 1997). Jordan et al (1994) using the polarisation-modulation FTIR and surface plasmon resonance spectroscopy proposed a model of the 11-mercaptoundecanoic acid (MUA)-PLy bilayer on the gold surface. The MUA molecules are oriented nearly perpendicular to the surface and the PLy backbone is aligned parallel to the gold substrates.

In our experiments we employed a similar approach, using carboxylic-rich PSMA for the immobilization of oligonucleotides/poly(L-lysine) complexes. A tentative scheme is presented in Figure 3. The feasibility of the reaction scheme was described elsewhere (Lin et al, 2002; Brian et al, 1996; Frey and Corn, 1996). Briefly, EDC and NHSS were used to covert the carboxylic acid groups of PSMA and chemically treated PSMAA to NHSS esters. The reaction of such activated surfaces with an aqueous solution of ε-amino groups of PLy will create an amide bond linkage with the polymer surface.

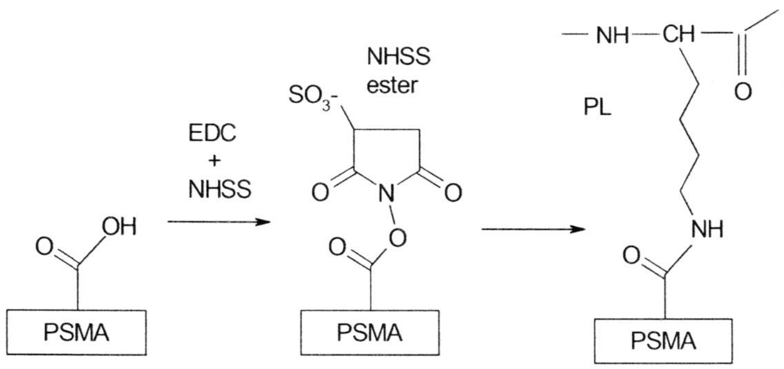


Figure 3. Surface modification scheme for attaching single stranded oligonucleotides onto PSMA/treated PSMAA via covalently bound PL. In the first step, the PSMA carboxylic acid groups react with EDC and then with NHSS to form the NHSS ester. In the second step PLy attaches to this intermediate by formation of amide bond.

The results obtained indicated that the employment of Ply clearly increased the immobilization efficiency of single-stranded DNA. Approximately up to 57% and 62% of the target oligonucleotide were bound to polymeric surfaces via electrostatic adsorption of PLy and covalent bonding of PLy, respectively. These values are in a good agreement with those reported for oligonucleotides that were covalently attached on amino-functionalised glass substrates, for which immobilization efficiency of 40 to 50% (Jordan et al, 1997) and 60-80% (Ghosh & Musso, 1987) under optimal conditions have been reported. On the other hand, the estimated levels of covalently bound DNA to activated Poly-styrene have reached 25 to 30 ng DNA per well (from 500-1.95 ng of bulk DNA solutions) (Rasmussen et al, 1991; Zammatteo et al, 1996, 1997). This is explained by the limited amount of the functional groups available at the surface which controls the amount of immobilized oligonucleotides, while Ply-modified surfaces provide higher density of binding sites and therefore increase the immobilization efficiency (Zammatteo et al, 2000).

The AFM analysis (Figure 1) of the DNA bound on PLy-mediated PSMA via covalent binding (a medium concentration case with regard to PLy surface concentration, as inferred from Table 1) confirmed high oligonucleotide yields on the surfaces tested and revealed that the PLy/DNA complex dramatically increases the rugosity of the surface. As opposed to the bare PSMA surface that presents only isolated defects, the PLy/DNA complexes aggregate in lumps of approximately 250 nm in height throughout the polymer surface. This structure of the film is preserved regardless of the zoom level, i.e. for 30 x 30 μm, or for 5 x 5 μm regions.

## Oligonucleotides Immobilization on Plasma Treated COC and PC Surfaces

Both the non-contact and contact AFM imaging showed that the covalently bound oligonucleotides were present as 'aggregates' on both NHS-functionalized polymer surfaces. The similar surface morphology obtained by the two imaging modes suggests that the oligonucleotides were strongly bound to the polymer surface (the contact mode 'drags' the tip along the sample, as opposed to the much gentler tapping mode). The frictional force data indicated that the oligonucleotide aggregates had higher frictional force, and as such, were more hydrophilic than the surrounding region. Figures 4 shows typical AFM topographical and corresponding frictional images of oligonucleotides immobilized on the COC and PC surfaces, respectively.

AFM analysis revealed that the oligonucleotide aggregates follow the topography of the respective surfaces, i.e. hill-like and terraced features on COC and PC, respectively. The aggregates on the PC surface had sizes typically in the range 0.2-0.8 μm, with heights of approximately 30 nm, with a surface coverage of 17%. A volume of the alu nucleotide at approximately 14 nm (as calculated using QSAR techniques as implemented in HyperChem) translates in an estimated density of $3.3 \times 10^{12}$ molecules/cm$^2$. Compared to the PC surface, the oligonucleotide aggregates on the COC surface were of a larger size, ranging from 0.8 to 2.0 μm, with heights of about 60 nm, with a much larger surface coverage of 41% and a density estimated to be $15.8 \times 10^{12}$ molecules/cm$^2$. Oligonucleotide coverage is influenced by a number of factors including the amount of available surface NHS groups, and the surface roughness. XPS data in this study indicated that the increase in surface oxygen following the plasma treatment is higher for the COC samples (Pham et al, 2002) suggesting that the NHS-functionalization, hence oligonucleotide binding, is more effective. The higher surface roughness of the NHS-functionalized COC surface may also increase the apparent oligonucleotide density. The binding kinetics of analyte molecules in solution to receptors immobilized on a surface has been shown to increase with surface heterogeneity and roughness (Sadana, 1998). In addition, surface density and surface diffusion coefficients of biomolecules including DNA oligonucleotides (Chan et al, 1997) and proteins such as BSA (Tilton et al, 1990) and prothombin fragments (Huang et al, 1994) have been reported to have an inversely proportional relationship. It is possible that the greater roughness of the COC surface may decrease the surface diffusion of the oligonucleotides and enhance the adsorption/binding events. The increased number of immobilized molecules will act as obstacles to diffusing molecules, thereby increasing the interaction between oligonucleotides. This is consistent with the larger size, both laterally and vertically, of the oligonucleotide aggregates on the COC surface compared with those on the PC surface. The interaction between oligonucleotides is also expected to increase with the length of the strands. The rough surfaces and the long strands (26 bases) used in the present study are likely to be major factors responsible for the larger size of the oligonucleotide aggregates compared to those reported in the literature (Lenigk et al, 2001) shorter oligonucleotides (16 bases) were immobilized on functionalized silicon-wafer surfaces.

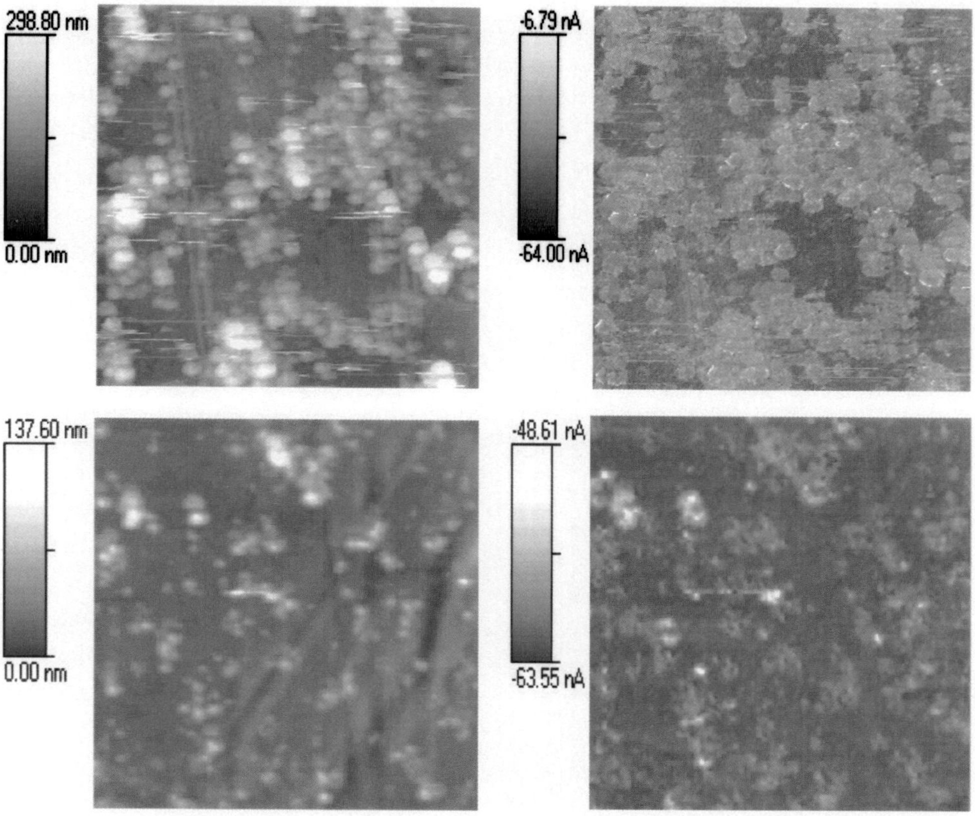


Figure 4. AFM topographical (left) and corresponding frictional (right) images of oligonucleotides immobilized on the COC (top row) and PC (bottom row) surfaces. The bright areas are more hydrophilic. Surface area in view is 20 x 20 μm.

# Complexation of DNA by Poly(L-lysine)

## Condensation of DNA by Poly(L-lysine)

In 1975 the mechanism of DNA condensation was dicovered to show that T4 and T7 bacteriophages pack their DNA by PLy condensation within their preformed heads. These condensates formed dough nut and stem-like structure as viewed under the electron microscope (Laemmli, 1975). Further experiments confirmed the existence of two types of these complexes. One was toroid/s with diameters averaging from 25 nm to 300 nm and another was rod/s shape with lengths of 40-80 nm (Bloomfield, 1996). It parallels the compaction of DNA by protamines inside the sperm head in which one turn of DNA wound around the circumference of the toroidal DNA structure of a radius of 50 nm is equivalent to 930 bp (Bloomfield, 1996). It has been also suggested that a rod shape condensed complexes can open up to form a toroid although the actual relationship between these two structures is yet to be clearly defined.

The condensation of DNA by PLy as a result of electrostatic interaction between the positively charged PLy and the negative charged phosphate groups in the DNA leading to electro-neutrality can easily be followed using agarose gel electrophoresis. The retardation of the ethidium bromide stained DNA band under ultraviolet light in increasing amounts of PLy indicates the reduction in negative charges left in the DNA complex. As the complex reach electro-neutrality not only it remains in the well but is no longer visible under ultraviolet light as a result of the compaction of the DNA complexed by PLy preventing the ethidium bromide chelation (Figure 5).

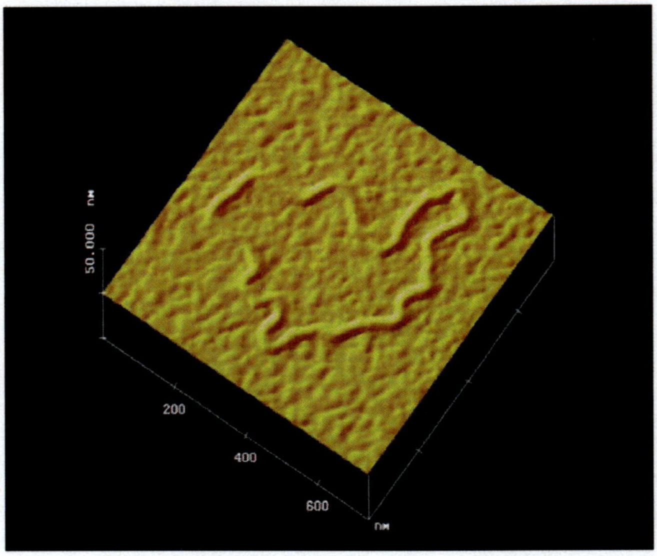


Figure 5. Gel retardation and ethidium bromide fluorescent quenching of pLy-DNA complex.

Other polyhomoamino acids including polyornithine, polyarginines also form complexes with DNA in water forming particles of sizes ranging from 80-120 nm. These particles were found to swell when they are further diluted into electrolytes and in cell media (Pouton et al, 1998).

## Complexation of DNA with polylysine of different molecular weights and in different buffers

The major factors that influence the formation of the different sizes, solubility and structures of the PLy /DNA complex include the charge ratio of the DNA to PLy which is defined by the moles of the amino group of the PLy divided by the moles of phosphate groups of the DNA, the size of the PLy itself, the modifications made to the PLy and the conditions in which the self assembly of the complexation occurs (Wagner et al, 1998; Ziady et al, 1999; Chan et al, 2000). Mannisto et al (2002) examining the condensing ability of DNA by PLy of different molecular weights showed that linear PLy of 2.9 kDa with 14 lysine residues binds to DNA very poorly while linear PLy of 20 kDa with 95 lysine residues binds DNA more efficiently at a charge ratio (+/-) of 0.6:1. Dendrimeric PLy generated in their lab were complexed with DNA and condensation compared with linear PLy. Small 3$^{rd}$ generation PLy dendrimers with molecular weight less than 1000 Da having only nine lysines are unable to bind DNA effectively while 5$^{th}$ generation PLy dendrimers although achieved effective DNA binding was found to be less effective than linear PLy at the same charge ratio (Mannisto et al, 2002). EM and AFM analysis of PLy of different lengths conjugated to the peptide ligand for the serpin enzyme complex receptor, found that shorter chain pLy (9.7 kDa)-DNA complexes were larger (AFM air dried samples: 39 +/- 2.5 nm) compared to the more compact pLy (53.7 kDa)-DNA complex of longer PLy chain ( AFM, 25 +/- 1.9 nm) as a result conferred less protection against DNase digestion (Ziady et al, 1999). Results showing increase in PLy /DNA complexation efficiency with increase in PLy length were further confirmed elsewhere (Hansma et al, 1998)

In the early 1970s using electron microscopy Haynes observed that DNA/PLy condensation resulted in different forms depending on the ionic strength of the buffer (Haynes et al, 1970). Using PLy of 4 kDa and 26 kDa, similar results were obtained by Hansma et al and the structures were analysed using Atomic force microscope. PLy was found to condense DNA poorly in neutral or alkaline saline even at high lysine:nucleotide ratios (Hansma et al, 1998). They were able to obtain complexes of PLy /DNA in water but binding to mica after air drying was poor. This was improved when the sample was diluted 5-10X in 40 mM HEPES, 10 mM $Mg^{2+}$.

Increase in NaCl concentration was shown to also impact complexation of PLy/DNA. Fluorescence quenching experiments show a reduction in total fluorescence in saline (0.15 M NaCl) than in water indicative of weaker binding of PLy to DNA in saline (Kwoh et al, 1999, Bhattacharya & Mandal, 1997). Laser light scattering analysis found that effective diameters of PLy/DNA in NaCl (0.15 mM) increased for all charge ratios. There were attempts to generate unimolecular DNA complexes with PLy (Perales et al, 1994; Ferkol et al, 1996; Perales et al, 1997) using high salt concentration (0.5 M) to first obtain a turbid solution followed by dilution until a clear solution is obtained. Using such an approach relatively homogenous particles of 12-20 nm were obtained.

Analysis of Ply/DNA condensates indicated that with plasmid DNA or oligonucleotides that are smaller, it is more likely that several molecules are condensed into a single aggregated structure (Bloomfield, 1996). Shorter DNA is also less likely to form toroids as there is insufficient length to form a nucleus. Toroids as observed from condensation with

DNA of lengths from 400 to 56 kb usually have an outer diameter of 50 nm and an inner diameter of 15 nm. (Wagner et al, 1991; Chan et al, 2000).

## AFM Visualization of pLy/DNA

Atomic Force microscopy has among its many advantages the ability to visualize condensed DNA complexes in a three dimensional topographic manner at a sub-nanometric resolution. Not only can one measure the length and width but the thickness of the complex can be estimated as well (Hansma et al, 1998; Chan et al, 2000). The direct visualization of the sample provides valuable information of the substructures revealing patterns, heterogeneous irregularities and defects not available with other techniques. The sample that has been applied to mica, can be air dried and subsequently be imaged under dry gas. Alternatively the sample can be imaged in solution. This provides a means in which the biological sample can be analyzed in physiologically relevant conditions and at the same time eliminating strong capillary forces between probe and sample. This useful and preferred as some complexes and samples acquire different structures or become unstable in air. Recent advances in AFM imaging of biomaterials in their native environment not only allowed visualization at the single biomolecular level but one can capture them at work and manipulation of the individual molecule itself can be accomplished (Engel & Muller, 2000; Fotiadis et al, 2002).

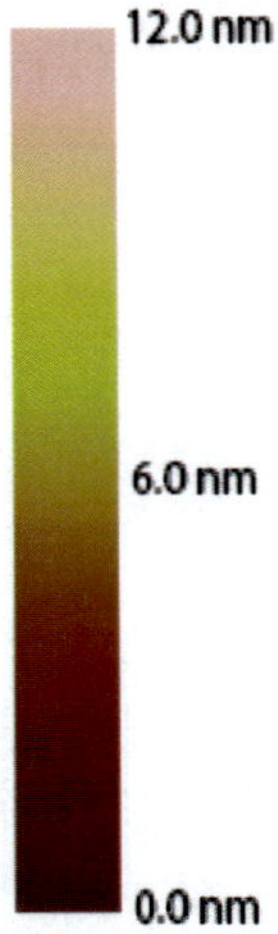


Figure 6. AFM visualization of an open plasmid. A 7.2 kb plasmid DNA in 40 mM HEPES and 10 mM MgCl2 pH7.6 (HBS) at a concentration of 10 ng/ul was added onto freshly cleaved mica, allowed to sit for 5 mins, washed with water and air dried with a stream of compressed air. The DNA was imaged in a tapping mode in air. The image showing the open plasmid aid is presented at an angle to emphasize topography.

AFM visualization in our lab of PLy/DNA complexes were carried out using a MultiMode AFM and Nanoscope III (Digital Instruments) performed on a tapping mode scanning at a rate of 1-2 Hz with 125 um silicon cantilevers oscillating at 400 Hz. Image of an opened 7.2 kb plasmid, allowed measurement of the thickness and the contours of the structure indicating that plasmid DNA had thickness of 0.32 +/- 0.04 nm (Figure 6) comparable those obtained by Hansma et al of 0.28 +/- 0.05 nm (Hansma et al, 1998). Toroids obtained as a result of condensation of pLy and the 7.2 kb plasmid has an outer diameter of 62.5 +/- 8.9 nm and an inner diameter of 53.5 +/- 3.1 nm. The thickness of these toroids was found to be 3.08 +/- 0.96 nm indicating nearly a 10 fold increase suggesting multiple layers or bundling of DNA strands.

### Stage Like Complexation of PLy/DNA at Increasing PLy/Nu Ratio

Analysis of Ply/DNA complexation in HEPES buffered saline (40 mM HEPES, 10 mM $MgCl_2$, pH7.6) at increasing Ly/Nu ratio using AFM in our lab has revealed a multi stage collapse of the complex. Progressive change in the higher-order structure of PLy /DNA complexes as a result of increasing PLy /Nu ratio has been reported in a number of other studies (Wagner et al, 1991; Golan et al, 1999; Chan et al, 2000). Electron microscopy analysis as well as AFM showed a stepwise collapse of the complex, from an open plasmid first to form toroids or rods observed when the PLy/Nu was at 0.4. Toroids formed as a result of complexation of a plasmid of 7.2 kb with PLy in our experiments had a diameter of 60-80 nm comparable to previous reports. As the PLy/Nu ratio increased, we have found that aggregated matrixes were formed with sizes ranging from 150 to 800 nm. These aggregates formed at PLy/Nu of 2.0 appeared in as various irregular structures. However as the PLy/Nu ratio further increase, a collapse of these structures occurred giving rise compacted beads with diameters ranging from 12-20 nm. Such stage like progressive collapse of PLy DNA structures due to increased in PLy/Nu (Table 1) parallels that found by a salt dependent Ply/DNA complex formation carried out by Perales (Perales et al, 1994; Perales et al, 1997).

Such variation in size and higher order structures at different PLy/Nu, as predicted had a significant impact on their use for gene delivery; on the ability of the complexes to be imported into the cell through the cell plasma membrane and particularly into the nucleus as discussed below.

## Conclusions

The analysis of the immobilization of oligonucleotides on functionalized polymer surfaces, namely PSMA, cyclo-olefine copolymer (COC) and polycarbonate (PC) surfaces revealed that the strategies applied for oligonucleotides immobilization, i.e. covalent binding routes and/or poly-L lysine-mediated, are capable to achieve high density of DNA required by DNA microarrays applications. AFM analysis showed that the immobilization of oligonucleotides on polymeric surfaces produced large aggregates and that the vertical and lateral dimensions on COC were typically larger compared to those on the PC and PSMA surface. While the more effective NHS-functionalization of the COC surface is a likely reason for the difference in oligonucleotide immobilization, it is suggested that the surface roughness of the substrate also plays a critical role. In particular, the higher roughness of the

COC surface may decrease the surface diffusion of the oligonucleotides, and increase the adsorption/binding events. The AFM frictional studies showed that the oligonucleotide aggregates were more hydrophilic than the surrounding surface. The relatively high immobilization efficiency (3-15 x $10^{12}$ molecules / $cm^2$), coupled with the stable thermo-mechanical properties, makes these polymers promising candidates for DNA-DNA hybridization assays integrated with PCR applications.

Results indicate that physical properties and morphological changes of the complexes formed under different conditions as the PLy complex with the oligonucleotide will impact the quality of the data collected. This is particularly critical to ensure optimal hybridization. The use of liner or dendrimeric pLy of different molecular weights and other different experimental conditions will result in structures of different sizes, compactness and morphology which may aid or hinder hybridization. These factors has yet to be examined rigorously with short oligonucleotides used in microarray analysis.

## References

Akey, CW, Visualization of transport-related configurations of the nuclear pore transporter. *Biophys. J.,* 1990 58, 341-55.

Bhattacharya, S, Mandal, SS, Interaction of surfactants with DNA. Role of hydrophobicity and surface charge on intercalation and DNA melting. *Biochim Biophys Acta,* 1997 1323, 29-44.

Beake, BD; Ling, JSG; Leggett, GJ. Scanning force microscopy of bulk-filled Uniaxially oriented poly(ethylene terephthalate) films. *Polymer,* 2000 41, 2241-2248.

Beamson, G, Briggs, D. High Resolution XPS of Organic Polymers, The Scienta ESCA300 Database. *Wiley, Chichester,* 1992.

Belosludtsev, Y; Belosludtsev, I; Iverson, B; Lemeshko, S; Wiese, R; Hogan, M; Powdrill, T. Nearly instantaneous, cation-independent, high selectivity nucleic acid hybridization to DNA microarrays. *Biochem. Biophys. Res. Comm.,* 2001 282, 1263-1267.

Bergström, CH; Ruotoistenmäki, J; Aitola, ET; Seppälä, JV. Influence of the polymerization conditions on the impact properties of metallocene-catalyzed copolymers of ethylene and aryl-substituted norbornenes with low exo/endo ratios. *J. Appl. Polym. Sci.* 2000 77, 1108-1117.

Becker, H; Gartner, C. Polymer microfabrication methods for microfluidic analytical applications. *Electrophoresis,* 2000 21, 12-26.

Bloomfield, VA. Condensation of DNA by multivalent cations: considerations on mechanism. *Biopolymers,* 1991 31, 1471-1481.

Bloomfield, VA. DNA condensation. *Curr. Opin. Struct. Biol.,* 1996 6, 334-341.

Boncheva, M; Scheibler, L; Lincoln, P; Vogel, H; Akerman, B. Design of oligonucleotide arrays at interfaces. *Langmuir,* 1999 15, 4317-4320.

Brian, L; Frey, BL; Robert, M; Corn, RM. Covalent attachment and derivatization of poly(L-lysine) monolayers on gold surfaces as characterized by polarization-modulation FT-IR spectroscopy. *Anal. Chem.* 1996 68, 3187 -3193.

Bustamante, C; Rivetti, C. Visualizing protein-nucleic acid interactions on a large scale with the scanning force microscope. Annu. Rev. Biophys. Biomol. Struct., 1996 25, 395-429.

Chan, CK; Jans, DA. Using nuclear targeting signals to enhance non-viral gene transfer. Immunol. *Cell. Biol.,* 2002 80, 119-130.

Chan, CK; Senden, T; Jans, DA. Supramolecular structure and nuclear targeting efficiency determine the enhancement of transfection by modified polylysines. *Gene Ther.,* 2000 7, 1690-1697.

Chan, V, McKenzie, SE, Surrey, S, Fortina, P, Graves, DJ. Effect of hydrophobicity and electrostatics on adsorption and surface diffusion of DNA oligonucleotides at liquid/solid interfaces. *J. Coll. & Interf. Sci.* 1998 203, 197–207.

Chan, V; Graves, DJ; Fortina, P; McKenzie, SE. Adsorption and surface diffusion of DNA oligonucleotides at liquid/solid interfaces. *Langmuir* 1997 13, 320-329.

Chastain, J. Handbook of X-ray Photoelectron Spectroscopy. Perkin Elmer Corporation *Physical Electronics Division* U.S.A.1992.

Cheng, J; Shoffner, MA; Hvichia, GE; Kricka, LJ; Wilding, P. Chip PCR. II. Investigation of different PCR amplification systems in microfabricated silicon-glass chips. *Nucleic Acids Res.* 1996 24, 380-385.

Chiu, SK; Hsu, M; Ku, WC; Tu, CY; Tseng, YT; Lau, WK; Yan, RY; Ma, JT; Tzeng, CM. Synergistic effects of epoxy- and amine-silanes on microarray DNA immobilization and hybridization. *Biochem. J.,* 2003 374, 625-632.

Chu, PP; Huang, W-J; Chang, FC; Fan, SY. Conformational conversion and chain ordering in cyclo olefin copolymer. *Polym. Commun.* 2000 41, 401–404.

Delpech, MC; Coutinho, FMB; Habibe, MES. Viscometry study of ethylene-cyclic olefin copolymers. *Polym. Test.,* 2002 21, 411-415.

Dolan, PL; Wu, Y; Ista, LK; Metzenberg, RL; Nelson, MA; Lopez, GP. Robust and efficient synthetic method for forming DNA microarrays. *Nucleic Acids Res.,* 2001 29, E107-7.

Engel, A; Muller, DJ. Observing single biomolecules at work with the atomic force microscope. *Nat. Struct. Biol.,* 2000 7, 715-718.

Evdokimov, YM; Platonov, AL; Tikhonenko, AS; Varshavsky, YM. A compact form of double-stranded DNA in solution. *FEBS Lett.,* 1972 23, 180-184.

Feldherr, CM; Kallenbach, E; Schultz, N. Movement of a karyophilic protein through the nuclear pores of oocytes. *J. Cell. Biol.,* 1984 99, 2216-22.

Ferkol, T; Perales, JC; Mularo, F; Hanson, RW. Receptor-mediated gene transfer into macrophages. *Proc. Natl. Acad. Sci.* U S A, 1996 93, 101-105.

Fodor, SPA; Read, JL; Pirrung, MC; Stryer, L; Lu, AT; Solas, D. 1991 *Science,* 251, 767.

Fotiadis, D; Scheuring, S; Muller, SA; Engel, A; Muller, DJ. Imaging and manipulation of biological structures with the AFM. *Micron,* 2002 33, 385-397.

Frey, BL; Corn, RM. Covalent attachment and Derivatization of poly(L-lysine) monolayers an Gold surfaces as characterized by polarization-modulation FT-IR spectroscopy. *Anal. Chem.* 1996 68, 3187-3193.

Ghosh, SS; Musso, GF. Covalent attachment of oligonucleotides to solid supports. *Nucleic. Acids Res.* 1987 15, 5353-5372.

Golan, R; Pietrasanta, LI; Hsieh, W; Hansma, HG. DNA toroids: stages in condensation. *Biochemistry,* 1999 38, 14069-76.

Grafström, S; Neitezt, M; Hagen, T; Ackermann, J; Neumann, R; Probst, O; Wörtge, M. The role of topography and friction for the image contrast in lateral force microscopy. *Nanotechnology,* 1993 4, 143-151.

Hansma, HG; Golan, R; Hsieh, W; Lollo, CP; Mullen-Ley, P; Kwoh, D. DNA condensation for gene therapy as monitored by atomic force microscopy. *Nucleic Acids Res.,* 1998 26, 2481-7.

Haynes, M; Garrett, RA; Gratzer, WB. Structure of nucleic acid-poly base complexes. *Biochemistry,* 1970 9, 4410-6

Hähner, G; Marti, A; Spencer, ND. The influence of pH on friction between oxide surfaces in electrolytes studied with lateral force microscopy: application as a nanochemical imaging technique. *Tribol. Lett.,* 1997 3, 359-365.

Huang, Z; Pearce, KH; Thompson, NL. Translational diffusion of bovine prothrombin fragment 1 weakly-bound to supported planar membranes - measurement by total internal-reflection with fluorescence pattern photobleaching recovery. *Biophys. J.,* 1994 67, 1754-1766.

Hartley PG; Bailey AI; Luckham PF; Batts G. Non-specific interactions between heparin and poly-lysine surfaces. *Colloids Surf A,* 1999 77 191-198.

Ivanova, EP; Pham, DK; Wright, JP; Nicolau DV. Oligonucleotide/poly(l-lysine) complexes attachment on poly(styrene/maleic acid) and poly(styrene/maleic anhydride) polymeric surfaces. *Biosensors Bioelectronics,* 2004 19, 1363–1370

Ivanova, EP, Papiernik, M, Oliveira, AM, Grodzinski, P, Nicolau DV. Feasibility of polystyrene-derived polymeric surfaces for covalent attachment of the oligonucleotides. *Smart Mat. & Struct.,* 2002 11, 783-791.

Jordan, C E; Frey, BL; Kornguth, S; Corn, RM. Characterization of poly-L-lysine adsorption onto alkanethiol-modified gold surfaces with polarization-modulation Fourier transform infrared spectroscopy and surface plasmon resonance measurements. *Langmuir* 1994 10, 3642-3648.

Jordan, CE ; Frutos, AG ; Thiel, AJ ; Corn, RM. Surface plasmon resonance imaging measurements of DNA hybridization adsorption and streptavidin/DNA multiplayer formation at chemically modified gold surfaces. *Anal. Chem.* 1997 69, 4939-4947.

Koch, T; Jacobsen, N.; Fensholdt, J; Boas, U; Fenger, M; Jakobsen, MH. Photochemical immobilization of anthraquinone conjugated oligonucleotides and PCR amplicons on solid surfaces. *Bioconjugate Chem.,* 2000 11, 474-483.

Laemmli, UK. Characterization of DNA condensates induced by poly(ethylene oxide) and polylysine. *Proc. Natl. Acad. Sci.* U S A, 1975 72, 4288-92.

Lenigk, R; Carles, M; Ip, NY; Sucher, NJ. Surface characterization of a silicon-chip-based DNA microarray. *Langmuir,* 2001 17, 2497-2501.

Lenigk, R; Liu, RH; Athavale, M; Chen, ZJ; Ganser, D; Yang, JN; Rauch, C; Liu, YJ; Chan, B; Yu, HN; Ray, M; Marrero, R; Grodzinski, P. *Analytical Biochemistry,* 2002 311, 40-49.

Lin, YC; Huang, MY; Young, KC; Chang, TT; Wu, CY. A rapid micro-polymerase chain reaction system for Hepatitis C virus amplification. *Sensors Actuat. B – Chem.,* 2000 71, 2-8.

Ling, JSG; Leggett, GJ. Scanning force microscopy of poly(ethylene terephthalate) surfaces: comparison of SEM with SFM topographical, lateral force and force modulation data. *Polymer,* 1997 38, 2617-2625.

McCabe, KM; Zhang, Y-H; Huang, BL; Wagar, EA, McCabe, ERB. Bacterial species identification after DNA amplification with a universal primer pair. *Mol. Gen. Metabolism,* 1999 66, 205-211.

McGovern, M.E.; Thompson, M. Thiol functionalization of surfaces for biosensor development. *Can. J. Chem.,* 1999 77, 1678-1689.

Mannisto, M; Vanderkerken, S; Toncheva, V; Elomaa, M; Ruponen, M; Schacht, E; Urtti, A. Structure-activity relationships of poly(L-lysines): effects of pegylation and molecular shape on physicochemical and biological properties in gene delivery. *J. Control. Release,* 2002 83, 169-82.

Northrup, MA; Ching, MT; White, RM; Watson, RT. DNA Amplification in a Microfabricated Reaction Chamber. in: Proceedings of the 7th International Conference on Solid State Sensors and Actuators (Transducers'93), June 1993 Yokohama, Japan, 924-926.

Northrup, MA; Gonzalez, C; Hadley, D; Hills, RF; Landre, P; Lehew, S; Saiki, R; Sninsky, JJ; Watson, R; Watson Jr, R. A mems-based miniature DNA analysis system in: Proceedings of the 8th International Conference on Solid State Sensors and Actuators, and Eurosensors IX, June 1995 Stockholm, Sweden, 764-767.

Noy, A; Frisbie, CD; Rozsnyai, LF; Wrighton, MS; Lieber, CM. Chemical force microscopy - exploiting chemically-modified tips to quantify adhesion, friction, and functional-group distributions in molecular assemblies. *J. Am. Chem. Soc.,* 1995 117, 7943-7951.

Nuzzo, RG; Dubois, LH; Allara, DL. Fundamental studies of microscopic wetting on organic surfaces. 1. Formation and structural characterization of a self-consistent series of polyfunctional organic monolayers. *J. Am. Chem. Soc.* 1990 112, 558-562.

Park, S; Healy, KE. Nanoparticulate DNA Packaging Using Terpolymers of Poly(lysine-g-(lactide-b-ethylene glycol)). 2003. *Bioconjugate Chem.* 2003 14, 311-319.

Perales, JC; Ferkol, T; Molas, M; Hanson, RW. An evaluation of receptor-mediated gene transfer using synthetic DNA-ligand complexes. *Eur. J. Biochem.,* 1994 226, 255-266.

Perales, JC; Grossmann, GA; Molas, M; Liu, G; Ferkol, T; Harpst, J; Oda, H; Hanson, RW. Biochemical and functional characterization of DNA complexes capable of targeting genes to hepatocytes via the asialoglycoprotein receptor. *J. Biol. Chem.,* 1997 272, 7398-7407.

Pham, DK; Ivanova, EP; Wright, JP; Grodzinski, P; Lenigk, R; Nicolau, DV. Surface characterization of oligonucleotides immobilized on polymer surfaces. In: Biomedical Applications of Micro- and Nanoengineering. 2002 *SPIE Proc.,* 4937, 115-124.

Pouton, CW; Lucas, P; Thomas, BJ; Uduehi, AN; Milroy, DA; Moss, SH. Polycation-DNA complexes for gene delivery: a comparison of the biopharmaceutical properties of cationic polypeptides and cationic lipids. *J. Control. Release,* 1998 53, 289-299.

Qureshi, SJ; Porteous, DJ; Brookes, AJ. Alu-based vectors and splinkers more efficient and comprehensive for polymerase chain reaction amplification of human DNA from complex sources. *Genet. Anal. Biomol. Eng.,* 1994 11, 95-101.

Rasmussen, SR; Larsen, MR; Rasmussen, SE. Covalent immobilization of DNA onto polystyrene microwells: the molecules are only bound at the 5'end. *Anal. Biochem.* 1991 198, 138-142.

Rogers, Y-H; Jiang-Baucom, P; Huang, J-Z; Bogdanov, V; Andersen, S; Boyce-Jacino, MT. Immobilization of oligonucleotides onto a glass support via disulfide bonds: a method for preparation of DNA microarray. *Anal. Biochem.,* 1999 266, 23-30.

Sadana, A. An analysis of analyte-receptor binding kinetics for biosensor applications: influence of the fractal dimension on the binding rate coefficient. *Biosens. Bioelectron.,* 1998 13, 1127-1140.

Schena, M; Shalon, D; Davis, RW; Brown, PO. Quantitative monitoring of gene-expression patterns with a complementary-DNA microarray. *Science,* 1995 270, 467-470.

Steel, AB; Levicky, RL; Herne, TM; Tarlov, MJ. Immobilization of nucleic acids at solid surfaces: effects of oligonucleotide length on layer assembly. *Biophysical J.,* 2000 79, 975-981.

Tilton, RD; Robertson, CR; Gast, AP. Lateral diffusion of bovine serum-albumin adsorbed at the solid liquid interface *J. Colloid Interf. Sci.,* 1990 137, 192-203.

Trau, D; Lee, TMH; Lao, AIK; Lenigk, R; Hsing, IM; Ip, NY; Carles, MC; Sucher, NJ. Genotyping on a complementary metal oxide semiconductor silicon polymerase chain reaction chip with integrated DNA microarray. *Anal. Chem.,* 2002, 74, 5480-5488.

Wagner, E; Cotten, M; Mechtler, K; Kirlappos, H; Birnstiel, ML. DNA-binding transferrin conjugates as functional gene-delivery agents: synthesis by linkage of polylysine or ethidium homodimer to the transferrin carbohydrate moiety. *Bioconjug. Chem.,* 1991 2, 226-31.

Wagner, E; Ogris, M; Zauner, W. Polylysine-based transfection systems utilizing receptor-mediated delivery. *Adv. Drug Deliv. Rev.,* 1998 30, 97-113.

Wilbur, JL; Biebuyck, HA; MacDonald, JC; Whitesides, GM. Scanning force microscopies can image patterned self-assembled monolayers. *Langmuir,* 1995, 11, 825-831.

Wolfert, MA; Seymour, LW. Atomic force microscopic analysis of the influence of the molecular weight of poly(L)lysine on the size of polyelectrolyte complexes formed with DNA. *Gene The.r,* 1996 3, 269-73.

Woolley, AT; Hadley, D; Landre, PA; deMello, J; Mathies, RA; Northrup, MA. Functional integration of PCR amplification and capillary electrophoresis in a microfabricated DNA analysis device *Anal. Chem.,* 1996, 68, 4081-4086.

Wu, GY; Wu, CH. Receptor-mediated in vitro gene transformation by a soluble DNA carrier system. *J. Biol. Chem.,* 1987 262, 4429-32.

Yoon, BJ; Lenhoff, AM. Computation of the electrostatic interaction energy between a protein and a charged surface. *J. Phys. Chem.*1992 96, 3130-3134.

Zama, M; Ichimura, S. Different effects of polylysine and polyarginine on the transition to a condensed state of DNA in polyethyleneglycol/salt solution. *Biochim. Biophys. Acta,* 1975 414, 256-62.

Zammatteo, N; Girardeaux, C; Delforge, D; Pireaux, J-J; Remacle, R. Amination of polystyrene microwells: application to the covalent grafting of DNA probes for hybridization assays. *Anal. Biochem.,* 1996, 236, 85-94.

Zammatteo, N; Alexandre, I ; Ernest, I; Le, L; Brancart, F; Remacle, J. Comparison between microwell and bead supports for the detection of human cytomegalovirus amplicons by sandwich hybridization. *Anal Biochem.* 1997 253, 180-189.

Zammatteo, N; Jeanmart, L; Hamels, S; Courtois, S; Louette, P; Hevesi, L; Remacle, J. Comparison between different strategies of covalent attachment of DNA to glass surfaces to build DNA microarrays. *Anal Biochem.* 2000 280,143-150.

In: Trends in DNA Research
Editor: Corey R. Woods, pp. 165-172
ISBN 1-59454-635-5


*Chapter VIII*

# The Application of Adhesive Tape for Collection of Evidence for Forensic DNA Analysis

***Richard C. Li*[1, 2] *and Howard A. Harris*[3]**
[1] Forensic and Investigative Sciences,
[2]Department of Biology, School of Sciences,
Indiana University-Purdue University, Indianapolis, IN
[3] Forensic Science Program, School of Public Safety,
University of New Haven, West Haven, CT

## Abstract

Known exemplar samples of human DNA have traditionally been body fluids, such as blood and saliva. In each case, the presence of water increases the risk for bacterial growth which may degrade the DNA evidence. In this study, the authors have developed a method that employed a hydrophilic adhesive tape (HAT) for collecting DNA evidence. The HAT method was used to remove surface cells from relatively hairless areas on the body. The areas examined were ankle, arm, behind the ear, between fingers and back of the neck. The HAT was then dissolved in the extraction buffer. DNA typing was performed using the short tandem repeat (STR) analysis. Our results show that the samples collected from the ear give the best results with a success rate of 100%. All subjects tested by this method had known STR genotypes established from buccal swabs. The author's results suggest that the HAT method can be used as a less invasive method for collecting biological evidence for forensic DNA analysis. In addition, this collection method should reduce the risk of DNA degradation due to the moisture.

**Keywords:** forensic science, STR, DNA evidence collection, tape lifting

# Introduction

Biological materials such as skin cells are often adhered to the adhesive tape used to collect evidence. Examples include latent fingerprint and gun shot residue evidence. Recent forensic DNA analysis techniques are sensitive and the forensic DNA analysis has been successfully performed on the biological evidence collected by adhesive tape. Studies reported successful DNA typing from biological material adhered to adhesive tapes associated with the collection of fingerprints [Van Renterghem *et al.*, 2000; Wickenheiser, 2002; Zamir *et al.*, 2000] and gun shot residue [Torre & Gino, 1996; Zamir *et al.*, 2004]. In this study, we have employed hydrophilic adhesive tape (HAT) for collection of evidence for forensic DNA analysis.

Traditionally, body fluid samples have been used as DNA control samples. In the early days of Restriction Fragment Length Polymorphism (RFLP) analysis, blood was the material of choice [Adams *et al.*, 1991; Budowle *et al.*, 1995]. As the technology has evolved, buccal swabs have become an alternative method [Walsh *et al.*, 1992; Sweet *et al.*, 1997]. Both blood and the saliva associated with buccal samples have the disadvantage of being fluid based and therefore susceptible to degradation from bacterial contamination [Adams *et al.*, 1991]. A relatively new method for collection is the use of FTA paper [Burgoyne *et al.*, 1994]. FTA paper is an absorbent cellulose-based paper that contains chemicals to protect DNA molecules from nuclease degradation and preserve the paper from bacterial growth [Burgoyne, 1996]. However, the collection of body fluids is somewhat invasive of the donor's person and under most circumstances a court order is a constitutionally mandated requirement. Although the buccal swab is less invasive, it still requires the cooperation of the individual being sampled and is much more likely to be subjected to external contamination.

Both of the above mentioned techniques provide samples that must either be processed and/or dried before they can be safely stored. Blood or swabs are more susceptible to contamination than a dry technique because of the associated moisture. In this study, the use of HAT allows significantly less invasive sampling than even a buccal swab because it requires only pressing the sticky tape against the skin in a hairless location. By placing the two sticky sides together the sample is protected from contamination, moisture, and most other possible environmental insults. Additionally, the sampling process can be performed at the posterior position (behind of ear) of the individual and may lower the risk of personal injury by uncooperative individuals. A portion of this tape can be cutoff and dissolved directly in the extraction buffer. The resultant solution can be sampled for quantitative analysis of the nuclear DNA present and then an appropriate aliquot taken for amplification and subsequent DNA analysis. Our work was reported previously [Li & Harris, 2003].

# Methods

## Sampling Procedure

Surface cells from the body area to be sampled were collected with a piece of the HAT (Scotch$^{TM}$, Wave Solder Tape 5414; approximately 3 X 6 cm). The sampling procedure was performed by firmly pressing the tape to the skin followed by lifting it free of the skin. The pressing-lifting cycle was repeated for each sample (up to eight times) until the tape seemed to no longer adhere to the surface when it was pressed against the skin. The translucent adhesive tape clearly showed the area of the tape that had contacted the skin. The tape was then folded over onto itself and stored in a sealed plastic bag at room temperature until needed for DNA extraction. Gloves were worn to eliminate any contamination from the individual doing the sampling. All items (except gloves) used for sampling were cleaned with 70% ethanol solution prior to the sampling.

### DNA Extraction

A strip of HAT (between 1 X 1 and 1 X 2 cm) was cut for DNA extraction. This small tape portion was placed in a microcentrifuge tube (1.5 ml). DNA was extracted by using the Chelex® extraction method [Walsh *et al.*, 1991]. 200 µl of Chelex® 100 (Bio-Rad Laboratories) solution (Tris, pH 8.0, 1 mM; EDTA, 0.01 mM; Chelex® 100, 5%) and the tape-buffer solution was heated at 56 °C for 20 minutes and then at 100 °C for 8 minutes followed by vortexing. The DNA extracts were then stored at -20 °C. The DNA yield extracted from the ear area was estimated at approximately 10 ng/200 µl range.

### Short Tandem Repeat (STR) Analysis

1) Four loci multiplex analysis. Up to 27 µl (approximately 14 %) of the total extract (200 µl) was used for DNA amplification. The fragments at loci VWA [Kimpton *et al.*, 1992], THO1 [Polymeropoulos *et al.*, 1991], F13A1 [Hammond *et al.*, 1994], and FES [Hammond *et al.*, 1994] were amplified. Polymerase Chain Reaction was performed according to the procedures of Kimpton *et al.* [1994] and Applied Biosystems [2000].

2) The COfiler™ and Profiler Plus™ systems (Applied Biosystems). Polymerase Chain Reaction was performed according to the procedure of Applied Biosystems [2000]. The amplified product was analyzed using an ABI PRISM® 377 Genetic Analyzer (4% denatured polyacrylamide gels) and an ABI PRISM® 310 Genetic Analyzer respectively.

## Results and Discussion

### The Effect of Dissolved Tape Solution on PCR Reaction

An experiment was designed to determine the amount of tape which could be employed for collection of DNA and which would not inhibit PCR amplification. A square 1 x 1cm of the tape was cut and dissolved in 200 µl TE buffer (Tris, pH 8.0, 1 mM; EDTA, 0.01 mM). Amplification was carried out with an increasing volume of the tape solution (0, 10%, 20%, 30%, 40% and 50% respectively) added to the reaction mix (total 25 µl). The peak height of TPOX locus was analyzed. The results demonstrated that no reduction of peak height was observed at TPOX locus when adding the dissolved HAT solution in proportions of 10-50 % of the final volume of PCR reaction mix (see Figure 1). This result indicates that no inhibition of DNA amplification was observed from adding the dissolved tape solution in proportions of 10-50 % of the final volume. In fact, there appears to be a small increase in yield after amplification in the samples that had the tape solution added. Since the main concern was the possibility of inhibition by the tape, this small effect was not further explored at this time.

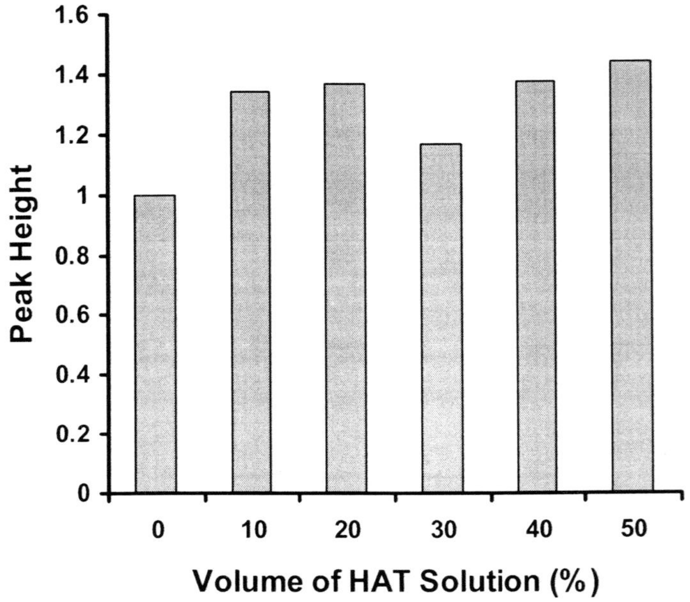


Figure 1. *The effect of dissolved HAT solution on PCR reaction.* Amplification was carried out with an increasing volume of the HAT solution 0, 10%, 20%, 30%, 40% and 50% respectively (25 µl final reaction mix). A 0.5 ng sample of human genomic DNA was used for template DNA. The peak height of TPOX locus was measured and was normalized for loading by measuring the peak height of ROX standard. Y-axis, the peak heights are expressed relative to those of the corresponding PCR reaction without HAT addition. The results shown are the averages of three experiments. The peak heights of TPOX locus were observed between 1446 to 3600 RFU (relative fluorescent unit).

## The Effect of Area of the Collection on DNA Typing

Surface cells from several different, relatively hairless, areas of the body were collected using the HAT method. The areas examined were 1) ankle, 2) arm inside elbow (crook of arm), 3) behind the ear, 4) between fingers, and 5) back of the neck. Our results show that the genotype profile analysis could be performed from the samples collected from the ear, finger, neck and arm. In contrast, no alleles were detected from the samples collected from the ankle area. The overall success rate for STR analysis was determined in terms of the percentage of alleles that were successfully detected. The success rate of four-loci multiplex system (VWA,

THO1, F13A1, and FES) is: ear, 100 % (13 subjects); finger, 65.5 % (13 subjects); neck, 58.3 % (5 subjects); and arm, 30.6 % (5 subjects) (Figure 2). Among the subjects tested for ear and finger, we included Caucasian (54%), Negroid (23%), Asian (23%), male (77%), and female (23%). The lower success rate observed in finger, neck, arm and ankle samples was due to the detecting of partial profiles or failure to detect a profile in some of the loci. This was mainly due to the occurrence of a low yield of DNA template and stochastic effect. We think that the amount of cells tape-lifted from these skin areas is lower than that of the ear area thus leads to a low yield of DNA template. A small trial (5 subjects) of ear sample was analyzed by using Profiler Plus$^{TM}$ system. The genotype profiles were successfully identified (Li & Harris, unpublished).

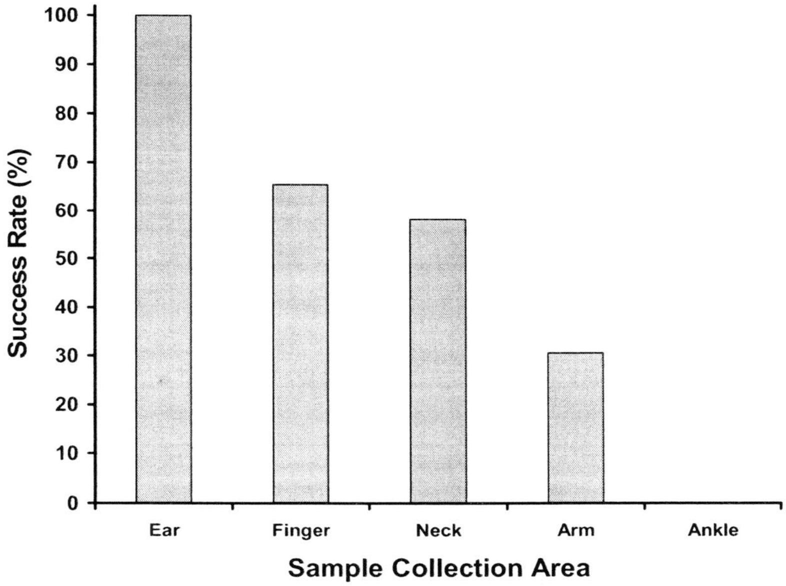


Figure 2. *The effect of area of the collection on STR analysis and comparison of success rate of STR analysis on four genetic loci.* The success rate is presented as the percentage of the number of alleles detected over the number of alleles tested. The success rates for STR analysis at four loci (VWA, THO1, F13A1, and FES) were compared among five body areas (ear, fingers, neck, arm and ankle). The success rate for STR analysis is presented as the percentage of the number of alleles detected over the number of alleles tested. Heterozygote was treated as 2 entries and homozygote was treated as a single entry. The success rates are: ear, 100% (13 subjects); fingers, 65.5% (13 subjects); neck, 58.3% (5 subjects); arm, 30.6% (5 subjects); ankle, 0 (5 subjects).

All subjects tested by the tape-lift method had known STR allele patterns established from buccal swabs. In all subjects, the STR allele patterns obtained from the ear samples (HAT method) corresponded to the known patterns generated from the buccal swab. One representative STR profile is shown in Figure 3 in which four loci (VWA, THO1, F13A1, and FES) were tested.

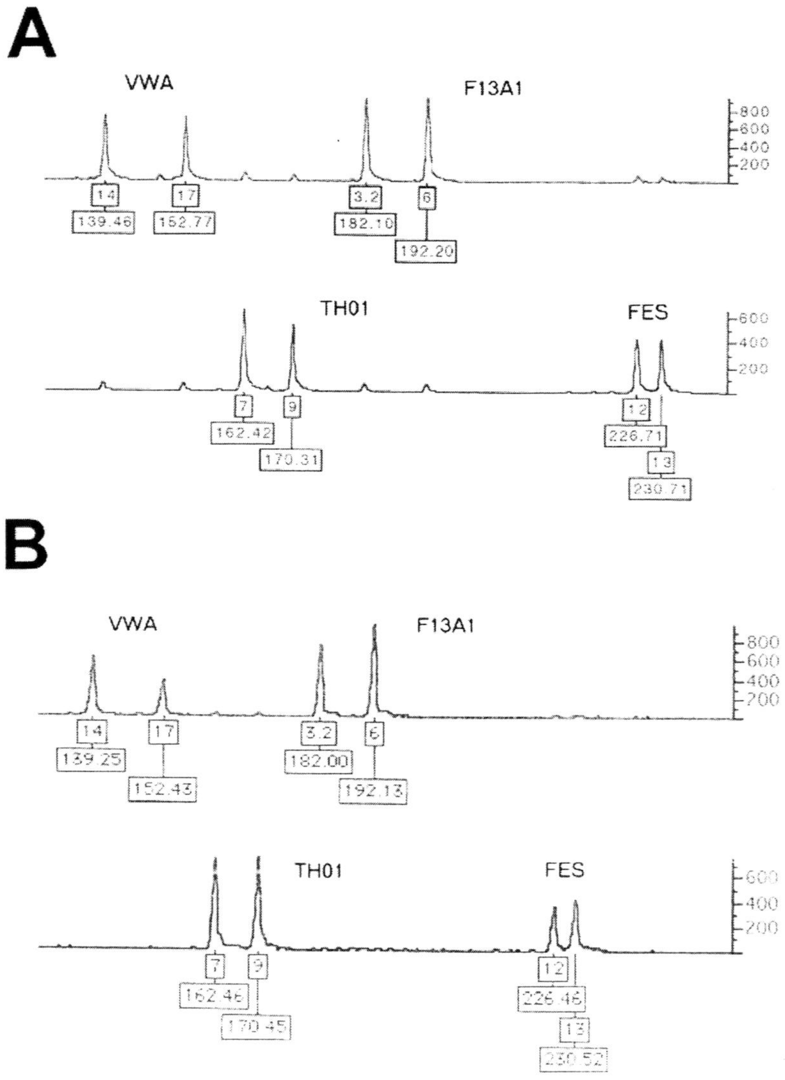


Figure 3. Comparison of DNA profiles of buccal swab samples and ear samples collected by HAT method. All subjects tested by HAT method had known STR allele patterns established from buccal swabs. Four loci (VWA, THO1, F13A1, and FES) were analyzed and compared. In all subjects, the STR allele patterns obtained from the ear samples (HAT method) corresponded to the known patterns generated from the buccal swab. One representative STR profile is shown. The Y-axis is expressed as RFU (relative fluorescent unit). A, buccal swab sample; B, ear sample collected by HAT method.

Results of this study show that the use of HAT provides a less invasive method for collecting DNA control samples and potentially evidence samples for STR analysis. The data demonstrates that DNA profiles can be obtained using HAT without an inhibitory effect on DNA amplification. STR analysis was successfully performed on the surface cells collected from several areas of the body using this HAT method. Also, the DNA genotype profiles obtained were consistent with those from other DNA sampling techniques such as buccal swab or blood collection. As indicated above, this collection method should reduce the risks of DNA degradation due to bacteria action, moisture and air, which are encountered using conventional collection methods. In addition, our results have demonstrated that when using the HAT method, the DNA samples are stable for one month after collection (Li & Harris, unpublished). The stability of the DNA samples under long-term storage should be studied in future experiments.

Recent studies reported DNA typing from biological material adhered to regular adhesive tapes. In our study, we reported a noninvasive method that employs a HAT method for efficient collection of DNA control samples. The HAT strip was completely dissolved during the DNA extraction process, thus we speculate that the yield of DNA obtained from using the HAT method should be higher than that of the regular tape method. The DNA analysis of evidence recovered from various substrates through skin contact has been reported [Wickenheiser, 2002]. The literature contains preliminary studies or case reports on the possibility of recovering DNA from articles such as garments, tools and weapons [Alessandrini *et al.*, 2003; Schulz & Reichert *et al.*, 2002; Lowe *et al.*, 2002]. Other potential applications of this HAT collection method will be explored. Developmental validation studies for possible evidence sample collection techniques will be conducted. The results of this study will be compared to that of other currently used techniques such as the swab method.

In conjunction with another application of the HAT, a holder was developed as a sampling devise. The holder allows the individual collecting the sample to put some pressure on the tape and to force it against the sampling surface. Use of such a device for sampling DNA from hairless body areas will be explored as part of this research as well as the collection of DNA materials from various articles.

## Acknowledgements

We would like to thank Larry Dow of ODV Incorporated for kindly providing HAT material used for this study.

## References

Adams, DE; Presley, LA; Baumstark, AL; Hensley, KW; Hill, AL; Anoe, KS; Campbell, PA; McLaughlin, CM; Budowle, B; Giusti, AM; Smerick, JB; Baechtel, FS. Deoxyribonucleic acid (DNA) analysis by restriction fragment length polymorphisms of blood and other body fluid stains subjected to contamination and environmental insults. *J Forensic Sci* 1991 36,1284-98.

Alessandrini, F; Cecati, M; Pesaresi, M; Turchi, C; Carle, F; Tagliabracci, A. Fingerprints as evidence for a genetic profile: morphological study on fingerprints and analysis of exogenous and individual factors affecting DNA typing. *J Forensic Sci* 2003 48, 586-92.

Applied Biosystems. AmpFlSTR Profiler Plus PCR amplification kit user's manual. Foster City, CA, Applied Biosystems 2000.

Budowle, B; Baechtel, FS; Comey, CT; Giusti, AM; Klevan, L. Simple protocols for typing forensic biological evidence: chemiluminescent detection for human DNA quantitation and restriction fragment length polymorphism (RFLP) analyses and manual typing of polymerase chain reaction (PCR) amplified polymorphisms. *Electrophoresis* 1995 16,1559-67.

Burgoyne, L; Kijas, J; Hallsworth, P; Turner, J. *Proceedings of the fifth international symposium on human identification*. Madison, WI. Promega Corporation 1994 163.

Burgoyne, L. Solid medium and method for DNA storage. *US Patent* 5, 496, 562. 1996.

Hammond, HA; Jin, L; Zhong, Y; Caskey, CT; Chakraborty, R. Evaluation of 13 short tandem repeat loci for use in personal identification applications. *Am J of Hum Genet* 1994 55, 175-89.

Kimpton, CP; Walton, A; Gill, P. A further tetranucleotide repeat polymorphism in the vWF gene. *Hum Mol Genet* 1992 1, 287.

Kimpton, CP; Fisher, D; Watson, S; Adams, M; Urquhart, A; Lygo, J; Gill, P. Evaluation of an automated DNA profiling system employing multiplex amplification of four tetrameric STR loci. *Int J Legal Med* 1994 109, 134-8.

Li, RC; Harris, HA. Using hydrophilic adhesive tape for collection of evidence for forensic DNA analysis. *J Forensic Sci* 2003 48, 1318-21.

Lowe, A; Murray, C; Whitaker, J; Tully, G; Gill, P. The propensity of individuals to deposit DNA and secondary transfer of low level DNA from individuals to inert surfaces. *Forensic Sci Int* 2002 129, 25-34.

Polymeropoulos, MH; Xiao, H; Rath, DS; Merril, CR. Tetranucleotide repeat polymorphism at the human tyrosine hydroxylase gene (TH). *Nucleic Acid Res* 1991 19, 3753.

Schulz, MM; Reichert, W. Archived or directly swabbed latent fingerprints as a DNA source for STR typing. *Forensic Sci Int* 2002 127, 128-30.

Sweet, D; Lorente, M; Lorente, JA; Valenzuela, A; Villanueva, E. An improved method to recover saliva from human skin: the double swab technique. *J Forensic Sci* 1997 42, 320-2.

Torre, C; Gino, S. Epidermal cells on stubs used for detection of GSR with SEM-EDX: analysis of DNA polymorphisms. *J Forensic Sci* 1996 41, 658-9.

Van Renterghem, P; Leonard, D; de Greef, C. Use of latent fingerprints as a source of DNA for genetic identification. In: Sensabaugh GF, Lincoln PJ, Olaisen B, editors. *Progress in Forensic Genetics*. Elsevier Science; 2000; Vol 8, 501-3.

Walsh, DJ; Corey, AC; Cotton, RW; Forman, L; Herrin, GL Jr; Word, CJ; Garner, DD. Isolation of deoxyribonucleic acid (DNA) from saliva and forensic science samples containing saliva. *J Forensic Sci* 1992 37, 387-95.

Walsh, PS; Metzger, DA; Higuchi, R. Chelex 100 as a medium for simple extraction of DNA for PCR-based typing from forensic material. *Biotechniques* 1991 10, 506-13.

Wickenheiser, RA. Trace DNA: a review, discussion of theory, and application of the transfer of trace quantities of DNA through skin contact. *J Forensic Sci* 2002 47,442-50.

Zamir, A; Oz, C; Wolf, E; Vinokurov, A; Glattstein, B. A possible source of reference DNA from archived treated adhesive lifters. *J Forensic Sci.* 2004 49, 68-70.

Zamir, A; Springer, E; Glattstein, B. Fingerprints and DNA: STR typing of DNA extracted from adhesive tape after processing for fingerprints. *J Forensic Sci* 2000 45, 687-8.

# Index

## A



## B

## C



## D

## H



## I



## J



## K



## L

## M



## N

## O



## P



## R

## S



## T

## U



## V



## W



## Y